Giant Molecules

Here, There, and Everywhere . . .

Giant Molecules

Here, There, and Everywhere...

Alexander Yu. Grosberg
Massachusetts Institute of Technology
Institute of Chemical Physics, Russian Academy of Sciences

Alexei R. Khokhlov
Moscow State University
Institute of Organoelement Compounds,
    Russian Academy of Sciences

Academic Press

San Diego   London   Boston
New York   Sydney   Tokyo   Toronto

Copyright © 1997 by Academic Press

ACADEMIC PRESS
525 B Street, Suite 1900, San Diego, CA 92101-4495, USA
1300 Boylston Street, Chestnut Hill, MA 02167, USA
http://www.apnet.com

Academic Press Limited
24–28 Oval Road, London NW1 7DX, UK
http://www.hbuk.co.uk/ap/

Library of Congress Cataloging-in-Publication Data

Grosberg, A. Yu.
    Giant molecules : here, there, and everywhere... / Alexander Yu. Grosberg and Alexei R. Khokhlov.
        p.   cm.
    Translated from an unpublished Russian manuscript.
    Includes bibliographical references.
    ISBN 0-12-304130-9
    1. Polymers.   I. Khokhlov, A. R.   II. Title.
QD381.G755   1997   96-27593
547.7—dc20   CIP

Printed in the United States of America
97 98 99 00 01 IC 9 8 7 6 5 4 3 2 1

Contents

9   Dynamics of Polymeric Fluids   157

Foreword
by P. G. de Gennes

The idea of atoms goes back to the Greeks: But for them it was really just a formal postulate, avoiding the intricacies of infinitely small objects. More than 2000 years were required to transform this into a reality, to show that the usual forms of matter around us are made of atoms and clumps of atoms that we call molecules. The first determination of the size of a molecule is probably due to Benjamin Franklin: knowing (again from the Greeks) that a small amount of oil suppresses the waves on the sea, he went to a pond in Clapham Common, choosing a day with a light wind, where the surface of the pond showed ripples. He then poured a spoonful of oil on the water, and measured the area upon which the ripples had disappeared: this area turned out to be huge. In our modern parlance, he had constructed a very thin monolayer of oil molecules. Dividing the volume (a spoonful) by the area, he could measure the size of a molecule (in this case, something like 2 nanometers).

Unfortunately, Franklin did not perform this calculation himself—it was done only a hundred years later by Lord Rayleigh (as explained in a beautiful book by C. Tanford[1]). But this experiment was a historical landmark: For the first time, molecules were not a figment of a philosopher's imagination. They became a physical object, with a well-defined number measuring their size!

A second step concerned the *giant* molecules that are the topic of this book. Many things around us (wood, cloth, food, our own bodies. . .) are made of macro-

---

[1] C. Tanford, *Ben Franklin Stilled the Waves*, Duke University Press, 1980.

molecules, or polymers as we call them now. But the concept of macromolecules emerged very slowly. During the 19th century, many chemists synthetized new polymers and threw them down the sink! In those days, the chemical dogma was to make a new substance, to purify it as much as possible, and to test the purity by measuring a property such as the melting point. If the melting point was sharp, the product was considered "good." But macromolecules, unfortunately, do not have a sharp melting point (for reasons that are, to a certain extent, explained in the present book). They were thus considered "dirty" and rejected. Ultimately, around 1920, H. Staudinger proved conclusively the existence of long chain molecules to the community of chemists. Physicists then entered the game: the first was Kuhn, who understood the flexibility of many polymers, the role of entropy in these systems, and the resulting elasticity of rubber. Again here, we have a great book describing the story.[2] Then came P. Flory, who mastered most of the physical properties of polymers, using very simple, but deep, ideas. The next step was due to S. F. Edwards, who pointed out a profound similarity between the conformation of a chain and the trajectory of a quantum mechanical particle. This allowed for fifty years of theoretical know-how accumulated in quantum physics to be transposed to polymer science!

The present book describes the final state of this evolution. The two Russian authors have a talent for writing it in a simple style—avoiding most of the heavy formalism that is beloved in countries of strong mathematical bias such as Russia or France.

The final product is accessible for university students and to research engineers. I am convinced that it will play a very useful role in this context. Giant molecules are important in our everyday life. But, as pointed out by the authors, they are also associated with a *culture*. What Bach did with the harpsichord, Kuhn and Flory did with polymers. We owe many thanks to those who now make this music accessible.

P. G. de Gennes
March 1996

---

[2] H. Morawetz, *Polymers: The Origins and Growth of a Science*, John Wiley, 1985.

(IMAGINARY) EDITOR   *(skeptically)*: Oh, not you again...

AUTHORS   *(bashfully)*: Well, you see, we've written a book on giant molecules...

EDITOR: *What* molecules?

AUTHORS: Giant ones *(getting more excited)*. Just listen to this bit here!

EDITOR *(impatiently)*: I haven't time to listen to it. Anyway, you've just published a book on those,[3] so what's the point?

AUTHORS: Yes, but that was for experts, and this one...

EDITOR *(losing his temper)*: And this one is for everyone else, presumably! Look, just leave the preface with me, and I'll see what I can do.

AUTHORS: Here it is!

Preface

For whom is this book intended?

We hope that this book will interest anyone with general curiosity about the world. It's not that we think so highly of ourselves; rather, what gives us this hope is the unique position of this field. It is right at the crossroads of so very many paths of contemporary development and ardent interest. Polymer physics

---

[3] A. Yu. Grosberg and A. R. Khokhlov, *Statistical Physics of Macromolecules*, AIP Press, 1994.

encompasses things from modern materials (including really fascinating "smart" materials) to the famous DNA, which is not just enthralling in its own right but is already becoming a tool that is used, for example, in criminology and as a "computer in a glass of water." Polymer physics is also about modern medicines, and lots more. In short, many things that people talk about every day have their roots in our science.

That is why we decided that it was time to write a clear, comprehensible story about giant molecules.

A college or university student should be able to read our book from cover to cover and get a superficial but coherent idea of the subject. A scientist—whether a physicist, chemist, materials engineer, or molecular biologist—may be interested to see how we discuss familiar topics while avoiding the complexities of scientific language. And finally, any reader may just browse through the book and find out, for example, what is meant by "molecular architecture," what happens when you chop up a cauliflower, and who used to be called the queen of the world and who her shadow was.

Just one more thing. In modern science, as in many other areas, aesthetic criteria are gaining more and more importance. In Russian, there is a well known saying by Dostoevsky: "Beauty will save the world." Of course, one can interpret these words in different ways, but there is no doubt that the beauty of science is one of its most astonishing features. Indeed, why does this most practical subject area happen to be most beautiful as well? We do not know, but it is a fact! So we have tried to demonstrate the beauty of this science.

Our two previous popular books on the physics of polymers[4] are still being read by both students and professors, not to mention various other types of readers. The first printing of our *Physics in the World of Polymers* was quickly sold out—71,000 copies altogether. For the present edition, we have written a couple of new chapters and sections, modified some old ones, and included many new figures.

Acknowledgments

We would like to thank Drs. M. A. Lifshitz and S. G. Starodubtsev, who read the manuscript and offered useful comments. We are indebted to Drs. T. A. Yurasova and C. J. B. Ford for their unlimited patience in the translation into English of

---

[4] A. Yu. Grosberg and A. R. Khokhlov, *Physics of Chain Molecules*, Moscow, Znanie, 1984; A. Yu. Grosberg and A. R. Khokhlov, *Physics in the World of Polymers*, Moscow, Nauka, 1989.

the text that we wrote in Russian. We are also obliged to Drs. S. V. Buldyrev and V. S. Pande for their help in preparing some figures and the CD ROM. Last but not least, we are deeply indebted to Professor Ilya M. Lifshitz (1917–1982); both of us were lucky enough to have him as a teacher. Quite a few of the scientific results we are going to discuss were originally obtained by him, and he was very good at creating a special atmosphere of ardent, inspiring interest in science. We have tried to recapture this atmosphere in our book, thus making it both interesting and useful. Whether we have succeeded or not is for our readers to decide.

Alexander Yu. Grosberg
Alexei R. Khokhlov
February 1996

EDITOR *(muttering to himself)*: Well, if they're not lying, perhaps it *is* interesting after all. It sounds like, apart from the general reader, the book may interest people in *(counting on his fingers)* the APS, ACS, MRS, . . . I think we ought to publish it!

Giant Molecules

Here, There, and Everywhere . . .

Introduction:
Physics in the World
of Giant Molecules

Molecules are supposed to be small, aren't they? Even the very word *molecule* comes from a Latin phrase that literally means "a tiny mass of something." Nevertheless, what would you say about a molecule approximately 1 meter long? Or another one that weighs almost 1 kilogram? There are many molecular giants of this kind, which are called polymers. Thus, our book is about polymers—and the world of polymers.

The world of polymers... Are polymers really so diverse and numerous that they make up a whole world? Is this not an exaggeration?

Well, what are polymers? The first things that come to mind may be plastic bags and other common plastics. You may also think of rubber and all its products. Then there are synthetic fibers and fabrics as well as natural ones, of course. In fact, the list is endless: cellulose (which makes up both timber and paper), the shell of a space probe traveling to Venus, artificial valves implanted into a human heart.... Polymers are used for all sorts of purposes, and huge quantities of them are made throughout the world. In fact, the volume of polymers produced already exceeds that of metals (although metals still win by weight).

The applications alone are a good enough reason to study polymers. However, it is not only their applications that make polymers so fascinating. The greatest incentive for polymer science is life itself. Even a schoolchild knows these days that our so-called "genetic blueprint" (i.e., what one is born to be—a dog or a cat, a boy or a girl—and what color of skin, hair, and eyes one is to have, etc.) is contained in molecules of a special polymer, DNA (deoxyribonucleic acid). Modern biology regards a living cell as a kind of factory, finely tuned, and controlled by DNA. Meanwhile, all the working devices in this factory (whether chemical, electrical, mechanical, optical, or whatever) are based on another type of polymer called proteins. In addition to this, polymers make hooves, horns, hair, and lots more!

It is not just that polymers are found in abundance in nature, they actually play a crucial role. So the Russian scientist M. D. Frank-Kamenetskii was not really joking when he called his popular book on DNA *The Most Important Molecule* (available in English with a much less impressive title; see reference [9] in the Suggested Readings at the end of the book).

Still not convinced? Let's just think how hard it has been for the most brilliant thinkers of humankind over hundreds of years, step by step, to unveil the mystery of life. Aristotle, Leonardo da Vinci, Darwin, Bohr, ... Yet it was not until the 20th century that knowledge in biology reached the molecular level. A key discovery was of the DNA double helix by James Watson and Francis Crick in 1953. All of a sudden surprisingly simple answers to the most fundamental questions started to take shape. And it has turned out that all these answers (yes, really, all of them!) are connected with polymers in one way or another.

You may say, "All right, I believe you, polymers are important. Perhaps one can even talk about the world of polymers if one wants. But why physics?" A good question that we shall answer shortly, but first, one more comment.

We would hate to sound like totally boring people who believe in doing only useful things. In fact, we believe it is a good idea sometimes just to pursue whatever takes your fancy! At least, that idea works very well in scientific research. After all, it is seldom clear from the start what use you can make of a discovery or idea. Fortunately, real scientists usually have good taste: What they like and want to do tends to also be useful.

So, let's go back to the question. Why study the *physics* of polymers? We can now give one good reason: It is very interesting! And it has a lot to offer, such as beautiful effects, fundamental analogies with other areas, and clear physical principles explaining complex phenomena. We shall try to give you a feel for the facets of polymer physics in this book. Of course, as for various applications, there are other people who can write a more complete story on each of those facets. Chemists could talk with confidence about synthetic polymers, and molecular biologists could expound biological polymers. However, even in these areas, physicists have no reason to feel too much out of place. Without physics, one can hardly reach a proper understanding of polymer chemistry or molecular biology. This is why all polymer scientists know the physics of polymers, and all use it to some extent in their work. Quite often the combination proves very fruitful.

There was even a period, in the 1940s and 1950s, when polymer physics was developed mainly by professional chemists. The most notable among them was P. Flory (1908–1982), an American physical chemist well known in scientific history for his pioneering work in polymer physics, for which he received a Nobel prize in 1974.

However, science is becoming more and more specialized. So it is not surprising that polymer physics has eventually grown into an independent field of research. This was helped by some eminent physicists, such as I. M. Lifshitz in Russia, S. F. Edwards in England, and P. G. de Gennes in France, who in the mid-1960s turned toward the study of polymers. These physicists revealed basic analogies between problems in polymer physics and some of the most burning and tantalizing questions of general physics. Polymers emerged on the pages of the world's main physics journals and at major international conferences. In 1991, P. G. de Gennes was awarded the Nobel prize in physics for his works. Rather rapidly, a harmonious system of simple models and qualitative ideas formed about the basic physical properties of polymers at a molecular level. All these concepts have been used successfully both in physical chemistry and in molecular biology. This also brought some terminology simplification. For example, we shall follow physics tradition and call the units of a polymer chain "monomers," not "monomer units" as chemists prefer.

In short, if you know about the physics of polymers you will understand why they are so widely used in everyday life and in industry, as well as how they work in biology.

 The book is accompanied by a CD ROM. There are several computer-generated movies on the CD ROM that illustrate various concepts and ideas discussed in the text. The program to generate these movies was written by Dr. S. Buldyrev. Although you can read the book independently of the CD-ROM movies, they can help to make many things more obvious. We have put special marks in the text, as here, every time we make comments related to the CD ROM. If you don't have a CD ROM, just disregard these places in

the book. The CD ROM contains not only the existing movies, but also the program engine, called application *Polymer*, which allows you to generate new movies. In fact, this is a unique tool for the curious mind, as virtually any process or phenomenon in molecular physics or chemistry can be modeled. Indeed, the possibilities of this instrument are far from being explored, much less exhausted, even by the authors. There are serious open scientific questions that can be addressed using application *Polymer*. In the Appendix at the end of the book, we explain how to use the program.

2

What Does a Polymer
Molecule Look Like?

2.1 — Polymers Are Long Molecular Chains

There was a time when in scientific essays all substances were described only in terms of how human senses perceived them. Even now one may come across this way of presenting things in some textbooks, as in "Water is a liquid that has no color, no taste, and no smell." These days such a description could also include information obtained from various measuring instruments, such as the spectrum or a material parameter. However, it would not be an exaggeration to say that modern scientists—be they physicists, chemists, or biologists—who study a substance should first of all have some image of a molecule of the substance.

Thus we shall start with chemical structure. Polymers are substances consisting of long molecular chains, so-called macromolecules. A helpful image is a long, entangled, three-dimensional thread, chain, rope, or wire.

What could be the chemical structure of such a macromolecule? Figure 2.1*a* shows schematically the structure of the simplest polymer chain, a polyethylene macromolecule. You can see that the macromolecule consists of CH_2 groups (called monomer units), which are connected by covalent chemical bonds to

5

FIGURE 2.1
Chemical structure of (*a*) polyethylene, (*b*) polystyrene, and (*c*) polyvinyl chloride.

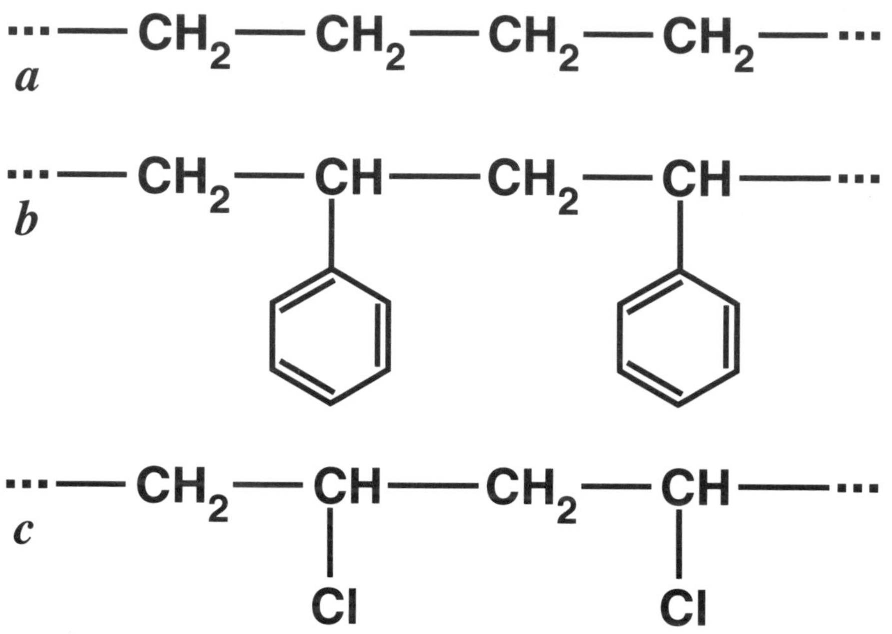

form a chain. Monomer units of other polymers (e.g., polystyrene or polyvinyl chloride) have different atomic structures (Figures 2.1*b* and 2.1*c*), but they are still organized into a chain of units.

To be considered a polymer, a molecule must consist of a great number of units, $N \gg 1$. Molecules of the types shown in Figure 2.1, if artificially synthesized in a chemical laboratory or industrial process, normally contain from hundreds to tens of thousands of units: $N \sim 10^2$–10^4. Natural polymer chains can be even longer than such synthetic polymers. The longest known polymers are DNA molecules; the number of monomer units in DNA can reach a billion ($N \sim 10^9$) or even ten billion ($N \sim 10^{10}$). It is just because they can be so long that polymer molecules are called macromolecules (*macro* is the Greek for large).

The fact that polymer molecules consist of long chains of monomers was not originally realized. At the beginning of the 20th century it was finally proved that matter consists of atoms and molecules. Yet no one attempted to look at polymers from a molecular point of view, even though some natural polymers (such as rubber, cellulose, silk, and wool) were widely used. At that time, the predominant opinion about polymers was that they were a sort of complex colloid system. It was not until the early 1920s that seminal works by the German physical chemist

H. Staudinger appeared. He suggested, after analyzing many experimental results, that polymer molecules are chains. The idea met with some scepticism at first—even with a fair amount of mockery in scientific circles. For instance, at one seminar Staudinger was asked, "So what kind of length are your molecules after all—the size of a nail, or of a finger?" All those present thought this notion very funny and burst into guffaws. Of course, from the modern point of view, there was nothing to joke about—DNA macromolecules, measured along the chain, can be as long as a few meters.

Even though his hypothesis was not accepted at once, Staudinger stuck to it and continued to accumulate more and more experimental evidence. As a result, by the beginning of the 1930s, the concept of the chain structure of macromolecules became generally established. It is sometimes reckoned that in the evolution of any scientific idea one can discern three different stages—at the beginning people say, "It's impossible!" then, "There may be something in it!" and eventually, "Oh well, but that's a well-known fact!" The concept that macromolecules are long molecular chains went through these three stages over a period of just ten years.

2.2 — Flexibility of Polymer Chains

The work by Staudinger prepared the ground for physics to intrude into the "polymer world." It had become possible to explain physical properties of various polymers by taking into account the chain structure of their constituents. But first, polymer scientists had to discern the specific shapes, or conformations, of molecular chains for different polymers.

For example, let's consider a polymer molecule diluted in some ordinary solvent (say, in water). What kind of shape does the molecule's chain have? Judging from the linear structure of the polymer chains (Figure 2.1), at first glance it seems reasonable to assume that the chain looks vaguely like a straight line (Figure 2.2a). But this is not true; as a matter of fact, it gets tangled up into a random, loose, three-dimensional coil (Figure 2.2b). This is simply a result of the chain's flexibility.

Generally speaking, the idea of flexible polymer chains may appear rather surprising. At school one is taught that the atoms in a molecule are joined together by covalent bonds in some specific order. Therefore their positions in space with respect to each other must be fixed too—just following from the formula for the structure. And if one looks at a small strand of the chain only, this argument will be absolutely correct.

FIGURE 2.2
(*a*) Rectilinear conformation of a polymer chain; (*b*) conformation of an entangled coil.

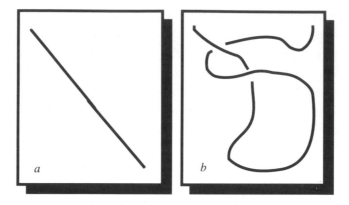

FIGURE 2.3
Spatial structure of a polyethylene chain segment in the most energetically favorable configuration.

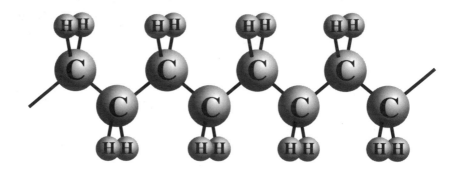

For example, Figure 2.3 shows the spatial structure of a little segment of a polyethylene macromolecule. You can see that the main chain is a sequence of carbon atoms connected with covalent bonds and that each carbon atom is also joined to two hydrogen atoms. So in complete agreement with the naive chemical concept, the atoms of each monomer unit as well as the atoms of neighboring units are located in a well-determined way with respect to each other.[1]

There is even a separate branch of research called conformational analysis of polymers. It deals with the geometry of atoms' positions in reasonably short chain segments (for much more complex structures than polyethylene, of course). An example is depicted in Plate 1; a strand of a DNA double helix. (We shall talk about DNA structure in more detail in Sections 4.5 and 4.6.) It is clear from the picture that each atom really does occupy a particular place.

---

[1] For the moment, we ignore the fact that the conformation of a polyethylene segment shown in Figure 2.3 is not the only possible one. A few different conformations can be realized because there are several rotational isomers of the molecule (we will discuss this later). By the way, this is the main reason for the flexibility of polyethylene chains.

Of course, the atoms of a molecule are not strictly fixed in their equilibrium positions. They may be pushed away from, and oscillate around, these positions due to various external effects, such as thermal collisions between a given macro-molecule and molecules of the solvent. In a real system, these oscillations hardly alter the lengths of covalent bonds. Changes in atoms' positions occur, first of all, because the bond angles (i.e., the angles between adjacent chemical bonds) can be deformed. In addition, parts of the molecule can rotate with respect to each other, around the axes of single covalent bonds (but not around double ones). This rotation is sometimes expressed in terms of a molecule having a few different "rotational-isomeric forms."

Thus, in many cases, you can regard a molecule as a construction of rigid rods, a bit like a miniature imitation of the Eifel tower. The rods swing slightly from side to side about the bonds. The amplitude of such bond-angle oscillations, as well as the probability of various rotational-isomeric forms, depends on the temperature. For example, at room temperature ($T \approx 300$ K) the oscillation amplitude of the bond angles ϕ for typical molecules normally varies from 1 to 10 degrees: $(\Delta\phi)_{T=300\ K} \sim 1$–10. Obviously, for an ordinary small molecule such oscillations would not appear too significant. Similarly, in a short segment of a polymer chain only low-amplitude fluctuations occur. This is why the chain's flexibility is hardly noticeable at such a small scale, and short chain segments can indeed be depicted in the way shown in Figure 2.3 and Plate 1.

At larger scales, however, all the small rotations add up along the chain and eventually result in the chaotic coiling of the polymer (Figure 2.2*b*).

 This is illustrated by the special movie called *"Flexibility"* on the CD ROM. Even if one starts with a polymer in an extended straight-line conformation, it is not too long before the polymer gets coiled up, completely losing any memory of its initial shape.

2.3 – Flexibility Mechanisms

As we have seen, any sufficiently long molecular chain does indeed have some flexibility, just because of its linear structure and considerable length. However, the nature of this flexibility may be different for different kinds of polymers. For example, the majority of the most commonly used synthetic polymers (including all those in Figure 2.1), as well as all protein molecules, have single C — C chem-ical bonds along their main chains. Such molecules appear flexible basically due to rotational isomerism, that is, because parts of a molecule may rotate around the

FIGURE 2.4
A wormlike polymer
chain.

single bonds. The main contribution to the discovery and study of this type of polymer flexibility was made by the physicist M. V. Volkenstein and his group from St. Petersburg (at that time Leningrad).

A classic example of a polymer with a different flexibility mechanism is a DNA double helix (Plate 1). Since it consists of two entwined "threads," rotations in one of them are prevented by the other. So the only remaining way in which the chain can flex is by deformation of the angles between the bonds. Each bond gets distorted slightly, so the flexibility is distributed fairly uniformly along the double helix. Nowhere may there be a kink or a right-angle bend, for example. DNA therefore looks like an elastic, wormlike thread, as shown in Figure 2.4. In fact, the model chain in Figure 2.4 is called a wormlike chain and is used to describe flexibility of this sort.

Yet another, and maybe the simplest, model for a polymer's flexibility is the so-called freely jointed chain. This is a sequence of rigid rods, each of length ℓ, joined together with freely rotating hinges (Figure 2.5). Such motions hardly ever

FIGURE 2.5
A freely jointed
polymer chain.

occur in a real polymer. However, as long as one is interested only in large-scale properties of a polymer coil, the particular nature of the chain flexibility ceases to be important. (We are going to discuss why this is so in Section 5.5.) Therefore, for the sake of simplicity, we shall use the freely jointed model in this book to explain some concepts and results.

2.4 — A Portrait of a Polymer Chain

A typical conformation of a freely jointed chain consisting of a great number of units is shown in Figure 2.6. You can easily create a similar pattern yourself; if you have access to a personal computer, it is also a good exercise in programming! However, if you only have a sheet of paper, we suggest the following routine. Draw a straight line of unit length. Then choose some random direction; you can do this, for example, by depicting a kind of a "wind-rose" (i.e., a diagram of the relative frequency of wind directions at a place) with six directions, numbering them in order from 1 to 6, and then tossing a die. (On a computer, instead of a die you would simply use a random number generator.) Now, starting at the end of your straight line, draw a new one of the same length in the chosen direction, and repeat this operation many times (i.e., choose a random direction again,

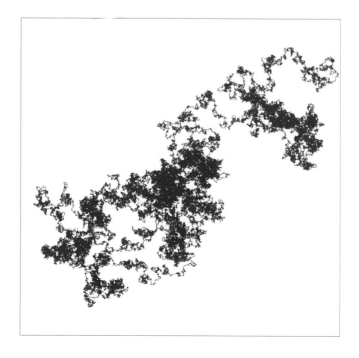

FIGURE 2.6
A typical conformation of a polymer coil. The freely jointed chain of 10^6 segments has been simulated computationally in three-dimensional space. The two-dimensional projection shown could have appeared as a Gaussian random walk on the plane, except the length of each step is not constrained to be unity. *Source:* Courtesy of S. Buldyrev.

independently from the previous one, add another straight line, etc). As a result, you get a "portrait" of a polymer chain, just like the one in Figure 2.6. Actually, this figure was obtained by a very similar procedure (on a computer); the only difference is that the wind-rose had many more than six different directions, and it was situated in three-dimensional space rather than on a plane.

Looking at Figure 2.6, you might think that you have already seen something similar when studying molecular physics. You would not be wrong, although there is no chapter on polymers yet in most textbooks on molecular physics. However, Brownian motion is included in all of them. They often show a photograph, made with a microscope, of the random path of a tiny dust particle suspended in a fluid and buffeted chaotically by numerous molecules. Such a random walk and the polymer conformation in Figure 2.6 are as alike as two peas in a pod. Why should this be the case? We are going to find out in Chapter 5.

Figure 2.6 also makes it clearer how a polymer chain tangles up into a random coil due to its flexibility (as we have already discussed; see. Figure 2.2b). You can reproduce the same kind of pattern using any model for a chain's flexibility; it does not have to be a freely jointed one.

Having read all this, you may be wondering why at the very beginning of our polymer story we are talking about such things as bending and the shape of polymer chains. In fact, bending of chains (or, in other words, their conformation) plays a key role in the properties of polymers. Nearly all of this book is a collection of examples of this, but here we shall give only one simple illustration. DNA molecules in human chromosomes are about a meter long. (There is quite a lot to be recorded there, hence the considerable length!) If DNA chains were not flexible but rigid like spokes, how could they be packed and kept in a cell nucleus as small as 10^{-6} m? As Plate 2 shows, this is the problem even for a bacteria.

2.5 — Heteropolymers, Branched Polymers, and Charged Polymers

You now know that what is special about polymers is their chain structure, great length, and flexibility. These are common features of all polymers, but they cannot explain everything. One complication is that each monomer unit has a particular chemical structure. Besides that, there are three major physical facts that make things more intricate, as we shall now discuss.

Heteropolymers Simple polymer chains, such as the ones in Figure 2.1, consist entirely of identical monomer units and are sometimes referred to as homopolymers. However, some macromolecules are built of monomer units of a few different sorts. They are known as heteropolymers, or copolymers as chemists say. Most interesting and important among them are biopolymers such as DNA (having four different types of monomer) and proteins (20 different types). The sequence of monomers along the chain forms the primary structure of this chain. One can compare the primary structure of a biopolymer to a sequence of letters in a very interesting and informative book written in a language that we do not yet completely understand.

Some heteropolymers are not biological, but are artificially synthesized. Their primary structures, in the spirit of our previous comparison, resemble a book that a monkey would have created if it were allowed to use a typewriter. Such a book would either be a totally random sequence of characters (i.e., a statistical copolymer) or a number of blocks of repeating identical letters, such as "BBBBBZZZCCCC" (i.e., a block copolymer). Lack of "sense" in their primary structures, by the way, does not prevent copolymers from having some very interesting physical properties or from being widely used in applications.

Branched Polymers   Together with simple linear chains, polymer science also deals with branched macromolecules. They can have the shape of combs (Figure 2.7a), stars (Figure 2.7b), or an even more complicated structure (Figure 2.7c).

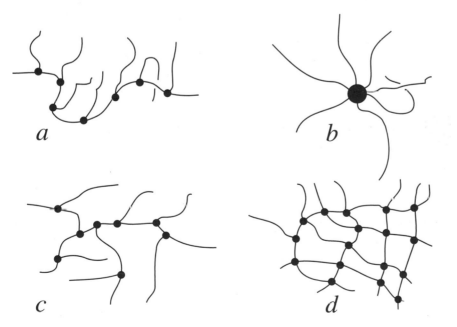

FIGURE 2.7
Branched macromolecules: (a) a comb, (b) a star, (c) a randomly branched chain, (d) a polymer network.

Another species of this kind is a macroscopic polymer network (Figure 2.7*d*) which takes the idea of branching to its extreme. This huge molecule emerges when numerous entangled polymer chains are chemically connected, or cross-linked, with each other (see Section 2.7). It can be many centimeters across and chemists have a special word for it: a gel. (Meanwhile, chefs, who may not even suspect that they are talking about polymer networks, use the same word in a slightly different form: jelly!)

Charged Polymers   None of the polymers depicted in Figure 2.1 contains electrically charged monomers. However, there are some polymers whose monomers may lose low molecular weight ions and become charged. Polymers of this sort are called polyelectrolytes, and the ions that break off are usually known as counterions.

The simplest of the polyelectrolytes are polyacrylic and polymethacrylic acids (Figure 2.8). When in solution in water, if an alkali is added the monomers of these polymers dissociate and become negatively charged. Biopolymers, such as DNA and proteins, also become polyelectrolytes when dissolved in water. DNA's chain has a large negative charge, whereas the monomers in proteins can be neutral or carry a positive or negative charge, depending on the type of monomer. By the way, such chains containing both positive and negative monomers are called polyampholytes.

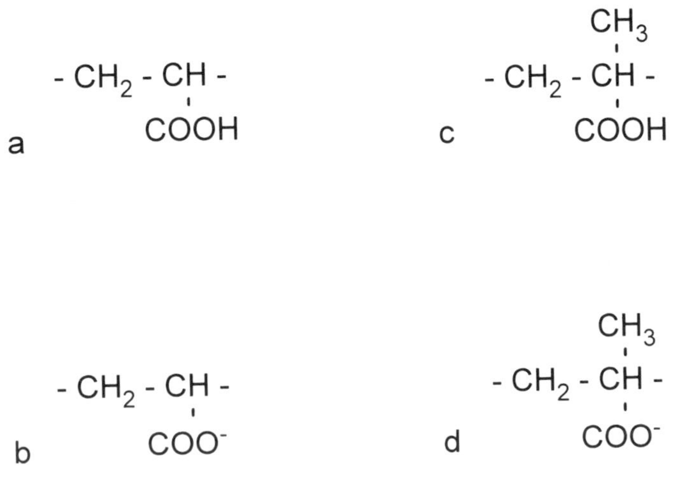

FIGURE 2.8
(a) A monomer unit of polyacrylic (*a*, *b*) and polymethacrylic (*c*, *d*) acids in the neutral (*a*, *c*) and charged (*b*, *d*) forms. The unit gets an electric charge by dissociation in water solution an alkali is added (e.g., NaOH; in this case the role of counterions for the charged units (*b*) and (*d*) is played by the Na$^+$ ions).

$$- CH_2 - CH -$$
$$a \qquad\qquad COOH$$

$$CH_3$$
$$- CH_2 - CH -$$
$$c \qquad\qquad COOH$$

$$- CH_2 - CH -$$
$$b \qquad\qquad COO^-$$

$$CH_3$$
$$- CH_2 - CH -$$
$$d \qquad\qquad COO^-$$

2.6 — Ring Macromolecules and Topological Effects

Some polymer molecules can have the shape of a ring (Figure 2.9*a*). Studying these, it is important to remember that parts of such closed chains cannot go through each other (Figure 2.10) in the way that ghosts, or phantoms, would do. In other words, as they sometimes say in scientific literature, the chains are "not phantom." Hence, the number of conformations in which a ring molecule can appear in its thermal motion is restricted. Anything that one can obtain from the original shape by various movements and deformations is allowed, but not the passing of the chain through itself. The mathematical properties of such objects are studied in a course on topology and are therefore called topological properties.

However, we do not need to know topology to understand that a ring molecule can be tied into a knot of some sort (Figure 2.9*b*). A few rings can form various entanglements with each other (Figure 2.9*c*). A peculiar thing about Figure 2.9*c* is that the molecules are not connected with chemical bonds, yet cannot be easily separated. Even such a thing as the so-called olympic gel (Figure 2.9*d*) can in principle exist. It looks like a kind of molecular chain-mail, and obviously acquired its name due to its resemblance to the coupled rings of the Olympic

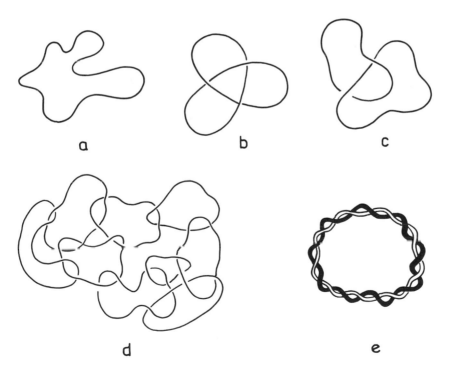

a   b   c

d   e

FIGURE 2.9
An unknotted (*a*) and knotted (*b*) ring macromolecule. (*c*) The tangling of two ring macromolecules. (*d*) An olympic gel. (*e*) The tangling of two complementary strands into a double helix.

FIGURE 2.10
An impossible type
of motion when
segments of a chain can
go through each other.

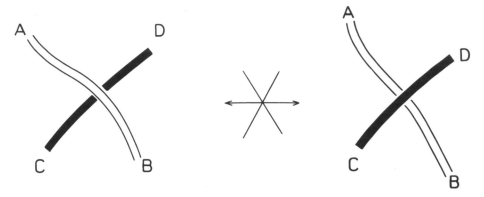

emblem. Of course, there are the same sort of topological constraints in polymer networks too (See Figure 2.7*d*).

One of the reasons why topological effects are of special interest is that natural DNA molecules normally, and perhaps even always, have a ring shape (Plate 8). The two strands of the double helix form a link of a very high order, as shown in Figure 2.9*e*. You may get some idea of how important the topology is from the following fact: Living cells have "provided" themselves with special topological enzymes that can do rather intricate jobs. They can, for instance, break one of the strands of a ring-shaped DNA molecule, then use some energy to "rearrange" the double helix by twisting it a particular extra number of times, and finally "heal" the break. Obviously, this is not just accidental, but is done for some good reason.

Linear polymer chains (of an open rather than a closed shape) are certainly not topologically constrained in the same sense. They can always come together or move apart. On the other hand, you have probably had to wrestle with a bundle of entangled ropes or cables. We all know how time consuming this is, and the knowledge that, in theory, ropes can easily be separated does not really help! So, based on this mundane experience, we may expect that systems of densely entangled linear chains should exhibit rather interesting and unusual dynamic behavior. We shall talk about this in Chapter 9.

2.7 — How Are Polymers Made?

We have talked about different types of molecules. Now let's explore how all these various types are actually made, ranging from the simplest linear polymer chain to a polymer network of a complex, densely entangled structure. In a living cell, chains of biopolymers are "built" by special systems in a process called biosynthesis, and there are also natural ways to prepare even more complex polymer systems.

But how are artificial polymers synthesized? This is a major task of polymer chemistry. This book, however, is meant to concentrate on physics, so we shall not discuss this question in any great detail. Nevertheless, it might help to have some general idea of the methods of polymer synthesis. It would let us understand physical properties of polymers better and more profoundly.

Long polymer chains are synthesized from low molecular weight compounds that are monomers. There are two main methods of synthesis: polymerization and polycondensation.

During polymerization, monomers are joined successively to the main chain, according to the rule $A_N + A \rightarrow A_{N+1}$. For example, polyethylene (Figure 2.1a) is obtained through polymerization of ethylene, $CH_2 = CH_2$. Under some conditions, two ethylene molecules can form butylene:

$$CH_2 = CH_2 + CH_2 = CH_2 \quad \rightarrow \quad CH_3 - CH_2 - CH = CH_2. \qquad (2.1)$$

Then a butylene molecule can react with another molecule of ethylene to make a six-carbon chain, and so on. "Capturing" more and more ethylene molecules, the chain becomes longer and longer and eventually grows into a macromolecule of polyethylene (Figure 2.1a).

From this example we can discern the main features of the polymerization process. First, to enable this kind of synthesis, a monomer molecule has to have at least one double (or triple) chemical bond. Second, the whole process is merely a rearrangement of chemical bonds between the molecules (e. g., a double bond transforms into two single ones). This is why no byproducts are normally created during polymerization, and the growing molecule consists of exactly the same atoms as the initial compounds.

It would be natural to ask here what conditions are needed for a chain to start growing. And when does the process stop? Apparently, a reaction like (2.1) cannot begin of its own accord. A lot of energy is required to create an active center of polymerization, which may be a free radical or an ion. The energy can come from heat, light, or radioactive radiation. Alternatively, one could sprinkle around some so-called initiators—special substances that can easily form free radicals (hydrogen peroxide is an example). As soon as an active center has formed, polymerization continues by itself with no outside help.

If an active center at the end of a chain ceases to exist (say, a free radical becomes a molecule, or an ion becomes an atom, etc.), then the chain stops growing. It is said in this case that a break in the chain occurs. This can happen for natural reasons (if, for example, two free radicals come together at the end of the chain), but it can also be deliberately stimulated by special substances called inhibitors. Obviously, the chain will also stop growing if the supply of monomers becomes exhausted.

Polycondensation is rather different. Segments of a polymer chain, with free radicals at the ends, gradually join on to each other: $A_N + A_M \rightarrow A_{N+M}$. "Esterification" is an example of such a process:

$$R-COOH + R'-OH \quad \leftrightarrow \quad RCOOR' + H_2O, \tag{2.2}$$

where R and R' are two segments of a polymer chain. The chain, of course, will grow further if the piece of the chain RCOOR' is able to attach itself to another similar piece; obviously, this can only happen if there are functional groups (like COOH or OH) at the ends of it.

During polycondensation low molecular weight substances are normally produced. (For instance, in the reaction (2.2) that substance happens to be water.) This is why, in contrast to polymerization, the content of a growing molecule changes compared to that of the initial compounds. Another special feature of polycondensation is the reversibility of reactions like (2.2). Creation of longer chains and their destruction are both happening at the same time. The latter, in fact, is mainly caused by low molecular weight products. So if one is aiming to obtain reasonably long polymer chains it is a good idea to get rid of the low molecular weight components during the reaction.

It would seem that polymer chains constructed from a mixture of monomers as a result of random chemical reactions should have a rather wide distribution in their lengths. This is indeed true, and the name for this phenomenon when chains of various lengths coexist in a polymer substance is polydispersity. Polydispersity has to be considered when analyzing polymer properties. In practice there are ways to reduce polydispersity by separating chains of different lengths.

Heteropolymers can also be synthesized as described above, but, of course, there must be a few different types of monomer in the mixture.

Now let's talk briefly about branched polymers. If, say, polycondensation is going on and the initial monomers have only two functional groups each, then we shall end up with linear polymer chains (with a small proportion of ring chains). However, if the monomers have three or more functional groups, a branched macromolecule can be synthesized (see Figure 2.7c). Given plenty of "multifunctional" monomer units at the start, one can even obtain a polymer network (Figure 2.7d).

Branched macromolecules and polymer networks can also be formed by the cross-linking of linear chains. There are various chemical ways of cross-linking. Sometimes chemically active cross-linking agents are used; they establish covalent bonds between different chain strands. Alternatively, ionization in a polymer system can be stimulated by radiation, etc. The simplest everyday-life example of cross-linking is vulcanization—during this process viscous natural rubber becomes a highly elastic polymer network (see Chapter 6 for more details).

3

What Kinds of Polymer Substances Are There?

Not from its separate parts, but from the sum of them, should a water pump be judged.
Koz'ma Prutkov
(satirical character in Russian literature known for his ostensibly "profound" statements)

3.1 — "Traditional" States of Matter and Polymers

We now can easily visualize polymer molecules—they are long chains tangled up into coils. However, even knowing the molecular structure of a substance, it may still be hard to predict for sure all its properties. For example, water (H_2O) can be a liquid, a solid (ice), or a gas (steam). So what about polymer substances? How do they look, and what states can they exist in?

**Apple pie, pudding, and pancake,
All begins with A.**
The Mother Goose Rhymes

Everyone is familiar with the three simplest ordinary states of matter: solid (crystal), liquid, and gaseous. Not everyone is aware that there is also a fourth one: a plasma. Normally it emerges at extremely high temperatures when thermal motion is so intense that it leads to the ionization of the molecules. If a polymer is heated up to such a temperature, its molecular chains will merely fall apart. At that point the substance is no longer polymeric; in other words, the destruction of the polymer occurs. Thus, the state of a high-temperature plasma is not possible for polymers.

It seems we are left with the three "traditional" states of matter after all. This sounds too small a number though, if we try to imagine all the diversity of polymer substances in everyday life: There are plastics and rubber, fibers and fabrics, timber and paper, polymer films and varnishes, dyes and paints—not to mention the various polymers found in nature! You would be right to suspect therefore that the common concept of the three states of matter is not quite applicable to polymers, especially since two of the three states, a gas and a crystal, are not really typical for polymers.

Indeed, if one wanted to create a polymer gas, one would have to make long heavy molecules (like the ones in Figure 2.7) "fly" around. This would only be possible if there were no gravity, and also if you could maintain a low pressure in the container (i.e., you would have to provide a high vacuum there). Obviously, such exotic conditions are very hard to achieve; this is exactly why polymer gases have not been heard of so far.

Perfect single crystals (see Figure 3.1) cannot be obtained from polymers for different reason. Let's experiment with a liquid of polymer molecules (Figure 3.1a). If we cool it down to below the crystallization temperatures, then the perfect crystal (Figure 3.1b) will be energetically the most favorable state. It cannot be formed straightaway, though. Crystallization goes on totally independently in different parts of the system. So what appears at the start is a number of crystalline "nuclei" randomly oriented with respect to each other. Clearly, when the nuclei grow big enough the entire structure becomes somewhat "frozen." (This is because, in the crystalline phase, to move with respect to each other polymer chains have to overcome enormous energetic barriers.) Hence, further evolution toward the perfect structure of Figure 3.1b appears hardly possible. This is why crystallizing polymers normally form a semicrystalline phase so that crystalline regions are separated by amorphous layers (Figure 3.1c). Sometimes perfect single crystals of a polymer can still be obtained by special techniques, but they have not found any broad practical use.

3.2 — Possible States of Polymer Substances

So, do we have to class all polymers as liquids, now we know that they can be neither gases nor, except rarely, crystals? In the broad sense, we would—if we only regard a liquid as a dense substance that has no long-scale order in the atoms' positions. However, this definition would not be terribly informative. This is why there is another, more fruitful way to classify polymers' phases. A distinction is normally made between a semicrystalline state, a polymer glass, an elastic, and

a

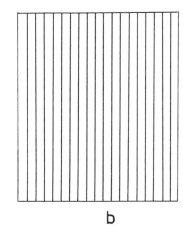

b

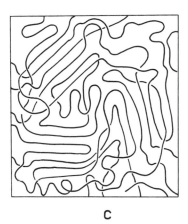

c

FIGURE 3.1
Polymeric structures:
(a) a liquid; (b) a perfect
crystal; (c) a partially
crystallized polymer.

a viscous polymer. Which of the four phases occurs depends on the kind and strength of interactions between the monomers.

We have already talked about semicrystalline polymers. Let's now describe in brief the other three states. A polymer in a viscous state is purely a liquid of macromolecules, as shown in Figure 3.1a. Long chains all mingle together, but, in thermal motion, they can rather easily move with respect to each other. If an external stress is applied, some overall motion of the molecules occurs; that is, the polymer starts to flow. The flow develops quite slowly, due to a great number of entanglements. This explains why the viscosity of polymeric liquids is normally rather high. Naturally, this state of a polymer is called viscous; another name for it is a polymer melt.

Let's now see what will happen if molecular chains of a polymer melt are joined together with covalent chemical bonds (cross-links) to form a network (see Figure 2.7*d*). (We talked about different techniques of how to synthesize a polymer network in Section 2.7.) Clearly, the chains will no longer be able to move long distances relative to each other (simply because they will all be tied together into a network). Thus it becomes impossible for the polymer to flow. Meanwhile, on a smaller scale (i.e., shorter than an average distance between two neighboring cross-links) the mobility of the chains will not be constrained by the cross-links. This is why, if you apply tension to a polymer network, its chains, which were originally coiled up (Figure 2.7*d*), stretch quite considerably, resulting in exceptionally large elastic reversible deformations. This state of a polymer is called elastic. Rubber is, obviously, a well-known example of it.

Cross-linking of chains in an elastic polymer does not necessarily have to be caused by covalent bonds between neighboring molecules, however. The role of effective cross-links can be performed by nuclei of a crystalline phase (Figure 3.1*c*) or by topological entanglements (Figures 2.9*c* and 2.9*d*). It can also be played by some small regions where, due to particular local conformations of the chains, there are comparatively high potential barriers for the chains to move with respect to each other ("glassy" or "frozen" regions). Thus, an elastic polymer substance can in principle be produced without chemical cross-linking.

If the temperature decreases, many polymers tend to change from a melt to a semi-crystalline state. However, far from all polymers crystallize when they are cooled. The crystal formation begins when little crystalline seeds start developing. This happens when, on one hand, the crystalline phase is thermodynamically favorable but, on the other hand, the thermal motion is enough to enable the rearrangement of the polymer chains to form seeds. If the cooling is fast enough, we can easily avoid that stage, and so a crystal does not form. This statement is also true for substances with low molecular weights. Polymers are special in that the "fast" cooling does not necessarily have to be very fast in the usual sense of the word. As you can see from Figure 3.1*c*, it takes much more time to form a crystalline seed for heavily entangled chains than for low molecular weight atoms and molecules that are not linked together.

Moreover, some polymers cannot be crystallized (in principle). Indeed, crystallization may only appear if there is long-range order in the molecules' positions (as in Figure 3.1*b*). However, say, for a statistical copolymer whose chains consist of two types of units, $\mathcal{A}$ and $\mathcal{B}$, long-range order is impossible. (This is simply because the sequences of $\mathcal{A}$ and $\mathcal{B}$ along the chains are totally random.) Such copolymers can never crystallize on cooling.

The same effect is observed for homopolymers whose monomers, although chemically identical, may appear in a few different spatial configurations. As an example, Figure 3.2 shows two possible configurations of the repeat unit

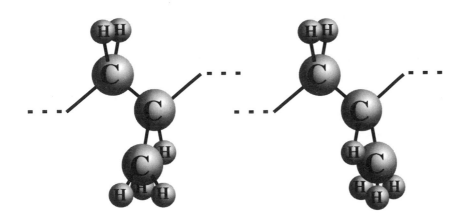

FIGURE 3.2
Two possible
configurations of the
monomer unit of
propylene.

of propylene. If the synthesis is carried out under usual conditions, these two configurations will be present in equal proportions and will alternate randomly along the chains. This kind of propylene is called atactic; obviously, it cannot crystallize. Yet there is a special technique for synthesizing the so-called isotactic propylene instead, whose monomers are all arranged in only one of the two possible configurations. Crystallization is then quite straightforward.

There are many other polymers that, like propylene, are atactic (i.e., unable to crystallize) if synthesized under normal conditions. This includes, for instance, polystyrene, polymethyl methacrylate (perspex), and polyvinyl chloride (PVC), to mention but a few from everyday life.

So what kind of processes *do* happen in the "noncrystallizable" polymers if the temperature is reduced? Since thermal motion becomes less, energy barriers for relative motion of the molecules grow higher and higher. Gradually, this motion becomes "frozen out," first of all at the largest scale, and then at increasingly smaller ones. In the end, any thermal motion at any scale larger than the size of a monomer ceases to exist. Polymers in such a "frozen" state are known as polymeric glasses, and the process that we have just described is called a glass transition. It normally occurs in a rather narrow temperature range around the "glass transition temperature" T_g.

Thus, polymers that are unable to crystallize tend to become a glass at low temperatures. You may be quite familiar with those more or less transparent glasses made from the atactic polymers already mentioned—polystyrene, perspex, and PVC. (For the first two of them $T_g \approx 100°C$, whereas for the last one $T_g \approx 80°C$.) However, ordinary silicate glasses (used for windows, for example) are not polymeric; they are made of low molecular weight compounds such as silicon dioxide and oxides of boron, sodium, and calcium. Such a mixture forms a glass on cooling, roughly in the same way as polymers.

The very term "glass" leaves no doubt that the substances are mainly used as transparent partitions. You may wonder why the majority of polymeric glasses are actually transparent, whereas semicrystalline polymers are normally not. The reason is that the structure shown in Figure 3.1c has a great number of interfaces between the crystalline and amorphous phases. Light gets reflected by the interfaces many, many times and eventually gets totally "lost" between them. As a result, the sample does not let light out, hence the lack of transparency. Meanwhile, acrylic glasses, say, have a much more uniform structure, or at least the scale of inhomogeneities is much smaller than the wavelengths of visible light. Therefore, light can penetrate through such substances without being scattered.

3.3 — Plastics

All four states of polymers are very important from the practical point of view. Rubber substances are used in their elastic state, for example. We shall talk about them in detail in Chapter 6.

In this section we are going to look at plastics, the materials that we know very well from everyday life. Over the last few decades they have become widely used in industry too. By definition, plastics are those polymers that while being produced are either elastic or viscous, whereas in actual use they have to be either glasses or semicrystalline substances. How can a polymer be transformed from one state into another? In many cases, it is done by changing the temperature: A polymer can be processed at elevated temperatures, where it is a viscous liquid, and afterward cooled to become glassy or semicrystalline. Materials produced in such a way are called thermosoftening plastics. However, some polymers tend to show the opposite behavior—they become solid with increasing temperature. For instance, epoxy resin mixed up with a hardener very quickly becomes solid if heated up. This is simply because cross-links are formed more rapidly at higher temperatures. Such materials are sometimes called thermosetting plastics.

Thus, all four states of polymers share "responsibility" for the properties of plastics. Some of them are involved at the production stage, and others come into play when the plastic is put to practical use. We should also point out in what way the properties of glassy and semicrystalline plastics differ from each other. Semicrystalline thermosoftening materials (such as polyethylene, terylene, nylon, and teflon) are much more deformable and elastic, and much less fragile, than polymeric glasses. Normally they are not transparent either, but, in contrast to rubber, they tend to retain their shape under moderate deformation.

The extent to which solid materials can be deformed is described in physics by Young's modulus, E. It is defined in the following way. Let us imagine that we are stretching a cylindrical rod of length ℓ and cross-sectional area S, applying a force f along the axis. As the English scientist Robert Hooke noticed as early as 1660, the deformation $\Delta\ell$ of the rod (i.e., variation of its length) is proportional to the force (provided that $\Delta\ell$ is not too big),

$$\sigma = \frac{f}{S} = E\frac{\Delta\ell}{\ell}. \tag{3.1}$$

In this formula σ is the stress; that is, it is the force per unit cross-sectional area, and E is Young's modulus. The value of E depends on the material of the rod, but not on its shape or size.

Let's now look at some materials that we are going to talk about later in this chapter and see what sort of values of E they have at room temperature. As a point of reference, it makes sense to choose the hardest inorganic substances, such as steel, cermet alloys, and so on. Their Young's moduli range from 10^{11} to 10^{12} Pa. Inorganic glasses (as used in windows) have E in the range of 10^{10} to 10^{11} Pa. For polymeric glasses, typical values are $E \sim 10^9 - 10^{10}$ Pa, which means that their deformability is two orders of magnitude higher than that of steel. As we have already said, semicrystalline plastics are even more easily deformed; indeed, they have $E \sim 10^8 - 10^9$ Pa. As for various sorts of rubber, as well as other polymers that are normally used in their elastic state, their Young's moduli tend to be exceptionally low: $E < 10^6$ Pa.

How can we account for such a great difference in values of E for different polymeric materials? Thermal motion in an elastic polymer is intense enough to enable the chains to move freely with respect to each other. However, long-distance movements of the chains (i.e., flow) are much harder to perform because of the cross-links. Under an external tension, the chains can be easily stretched—this explains very low values of Young's modulus.

In contrast, in polymeric glasses, relative motion of the chains is hardly possible even on scales as small as the size of a monomer. Therefore their Young's moduli are essentially higher. A detail to notice here is that for polymeric glasses room temperature values of E are still an order of magnitude lower than those for inorganic glasses. This shows that at room temperature motion is not as "frozen" in polymers as in silicate glasses; there is still some freedom for the chains to rearrange their conformations locally, which increases deformability and reduces the value of E.

Finally, in a semicrystalline polymer there are amorphous partitions between regions of crystalline phase (Figure 3.1c). For many materials these partitions are not completely in the glassy state at room temperature. So what we really have is a mixture of solid crystal "islets" separated by a kind of "grout" made from a rubber-like polymer. Clearly, such material should be less fragile, with lower values of Young's modulus, than a polymeric glass.

3.4 — Polymeric Fibers

We have described possible states of polymers and have discussed how the most commonly used materials, plastics and rubber, come into the picture. However, there is another class of polymeric materials that we have missed so far, namely, fibers. Fibers are by no means any less important, one good reason being that nearly all our clothes are produced from fibers. So what are they, and what state of matter do they represent?

First of all, we should note that polymeric fibers can be either natural or produced in a chemical laboratory or a factory. Cellulose, for instance, is the most widespread natural fiber. Molecular chains of cellulose form the walls of biological cells in most plants; they are also the chief constituent of timber. Natural cellulose fibers are obtained from flax, cotton, hemp, and so on. Other well-known kinds of natural fiber are wool and silk. They, of course, have "animal" origin: Wool is given to us by sheep, goats, and camels, whereas silk is produced by a caterpillar (silkworm) with the Latin name *Bombyx mori*. (It is amazing that a single thread of fiber made by a silkworm is about a kilometer long!) Chemically, wool and silk consist of polymer chains of particular proteins, called keratin (wool) and fibrion (silk).

The fibers obtained in a factory or a chemical laboratory are called chemical fibers. There are two types: artificial fibers that are made from natural threads but modified to improve their properties and synthetic fibers synthesized from some simple chemical compounds.

Artificial fibers are mainly obtained from various kinds of natural cellulose. For example, timber cellulose is used to make viscous rayon, and the so-called acetate and triacetate fibers can be produced from cotton cellulose.

Among the most interesting synthetic fibers are nylon and its various brands (called polyamide fibers because they are built from polyamide molecular chains), terylene (polyether fibers), and lastly orlon and acrylon (polyacryl nitrile fibers).

The particular physical state of polymeric fibers in which they are actually used depends on the purpose they serve. Obviously, the fibers have to be reasonably tough and should not stretch significantly under the influence of longitudinal forces that occur in the fiber during its use.[1] This immediately rules out the viscous and the elastic states. As for the other two states, the fibers can be either semicrystals (cellulose fibers, nylon, and terylene) or polymeric glasses (orlon). There is, however, something special about the structure of the fibers in these states.

Are semicrystalline fibers arranged in the same way as in Figure 2.1*c*, and polymeric glasses as in Figure 2.1*a*? In fact, if they were, such materials would be of rather poor quality. They would not be strong enough, but would be quite easily stretched (compare the values of the elastic modulus given above for typical polymers in the two states). Experiments on semicrystalline natural fibers have revealed, however, that their crystalline regions are not oriented randomly (as in Figure 2.1*c*), but are mainly parallel to the axis of the fiber (Figure 2.3*a*). This is just the kind of structure, with the chains predominantly parallel to the axis of the fiber, that is aimed at when chemical fibers are synthesized (both polymeric glasses and semicrystals, Figure 2.3*b*). Thus, polymer fibers are always anisotropic because polymer chains have a preferential orientation along the axis of the fiber. The higher the anisotropy, the greater Young's modulus for longitudinal deformations, and the stronger the fiber.

How can we explain this? Why should a fiber become stronger when its anisotropy increases? To answer this question, let's look once again at the perfect polymer crystal shown in Figure 3.1*b*. What happens if we start stretching the sample along the direction of the polymer chains' orientation? The stretching will be hindered by the covalent bonds that hold the monomers together in the long chains. Say the chains have carbon "spines" (which is true for many important polymers, e.g., polyethylene and polyvinyl chloride). In this case the covalent bonds in charge of building up the chains are obviously C—C bonds. We can estimate their deformability if we remember that the diamond, whose crystalline structure is also formed by C—C bonds, has Young's modulus $E \sim 10^{12}$ Pa. Naturally, for the crystal shown in Figure 3.1*b* the order of magnitude of E should be the same. The breaking strength for the two materials should be reasonably similar too (but only, of course, if the tension is applied to the sample in Figure 3.1*b* along the polymer chains).

On the other hand, we know that for disordered semicrystalline polymers $E \sim 10^8 - 10^9$ Pa because of the amorphous layers between the crystalline areas.

---

[1] By the way, for the fibers used in the making of clothes, the limiting factor is not the stress caused by the wearing of the clothes. When making cloth from fibers, the fibers are first spun into thread, and then the thread is woven or knitted to form the final product. In mass fabrication, stresses occur during the spinning and weaving or knitting stages. These stresses are normally much higher than those that arise during the use of the finished fabric.

If the crystalline areas start getting ordered in a particular direction (in other words, if the structure in Figure 3.1c starts transforming into the structure in Figure 3.3a), then more and more chains appear to be stretched along the axis of the fiber. Such chains take most of the strain arising from the deformation; they make the fibers considerably stronger, and the Young's modulus increases. Certainly, one cannot achieve the strength of diamond in such a way. Yet it is quite possible to improve the mechanical properties of the material by 1.5 − 2 orders of magnitude, due to an increase in the anisotropy of the fiber. Exactly the same idea is used to strengthen glassy fibers—making the transition from Figure 3.1c to Figure 3.3a.

So far we have talked about the final structure of fibers when ready for use. In the fibers obtained from natural products (e.g., from cotton or wool), nature itself has provided the right structure. Indeed, these fibers, though semicrystalline, are highly anisotropic (as in Figure 3.3a). But what about chemical fibers? You may

FIGURE 3.3
The structure of oriented
(a) semicrystalline and
(b) amorphous fibers.

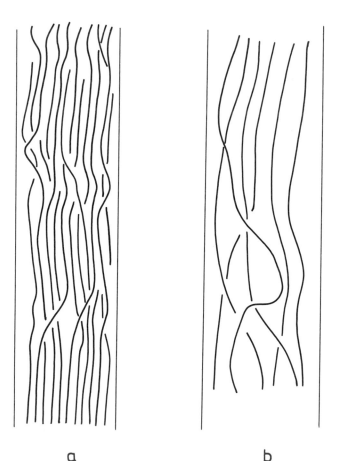

a   b

ask, first of all, how they are made, and, secondly, how the necessary degree of anisotropy is provided.

The usual strategy is as follows. Take a polymer from which you want to make a fiber, and convert it to a viscous state. This can be done by heating it up if the polymer has a reasonably low melting point, like nylon or terylene. Otherwise, for refractory polymers in particular, polymers can be diluted in a good solvent; the result is a concentrated fluid solution of the polymer, called a spinning solution. Then the fiber is made by squeezing the solution (or the melt) through little holes (spinnerets) into a medium that helps to solidify the polymer in the shape of thin fibers. If the fibers are formed from a melt, the medium can just be cold air. On the other hand, if a solution of a polymer is used, the solvent has to be removed from the fibers after passing through the spinnerets. The solvent can be evaporated by placing the fibers in a jet of hot air. Alternatively, the threads can be treated in a so-called precipitating bath, which contains a special medium that makes it energetically favorable for the polymer to shrink and squeeze out the solvent.

However, the fiber obtained in such a way is not yet sufficiently aligned; its structure looks like that in Figures 3.1a or 3.1c. To make structures like those in Figures 3.3a and 3.3b, one then has to stretch the solid fiber at a temperature high enough for the polymer not to form a glass. This process, called orientational stretching, causes the polymer chains become more aligned. Figure 3.4 shows a typical dependence of the stress σ on the relative elongation $\Delta\ell/\ell$ for fibers made from semicrystalline polymers. If σ is not too big, Hooke's law (equation 3.1) is valid; the deformation in this case is elastic (reversible) and the fiber starts going back to its initial state after the force has stopped acting. However, when the

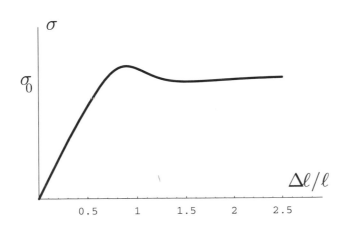

FIGURE 3.4
Typical dependence of stress σ on relative elongation $\Delta\ell/\ell$ for an isotropic semicrystalline fiber.

FIGURE 3.5
The development of a
"neck" during the
stretching of a fiber.

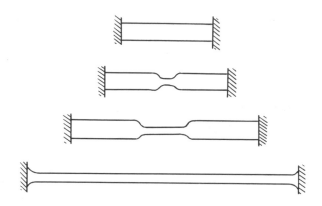

stress becomes as high as σ_0 (see Figure 3.4), the situation changes dramatically. The deformation starts increasing of its own accord whereas the stress remains the same or even decreases slightly. During this process a sort of a "neck" develops in the fiber; it lengthens and eventually runs the whole length of the sample (Figure 3.5). As a result, the fiber may stretch by a factor of about two. Naturally, the deformations occurring in the fiber after the "neck" has been formed are irreversible.

Here is a very simple experiment you can easily do yourself to show that the processes described really do take place in polymer fibers. Cut carefully a strip of polyethylene, 10 cm long and 1 cm wide. Stretching it, you will see that, after a certain amount of deformation, a very elongated zone appears in the middle of the strip. It spreads along the whole strip with further deformation. This zone is similar to the "neck" developing on a stretched polymer fiber—as soon as it emerges the deformation becomes irreversible.

At a molecular level, the development of the "neck" means that the stress applied to the fiber is apparently rather high; therefore it leads to the breakup of the semicrystalline structure with random orientation of the crystallites (i.e., the one shown in Figure 3.1c). In the "neck" zone, the structure rearranges since polymer chains align on stretching. As a result, the chains are eventually oriented as shown in Figure 3.3a and the fiber becomes anisotropic and therefore stronger.

The strengthening is also helped by one more thing. When Figure 3.1c is transformed into Figure 3.3a, the degree of crystallization increases (quantitatively, there is an increase in the volume fraction of crystalline regions in the semicrystalline fiber). This happens because some preliminary orientation of the chains paves the way for further crystallization. The latter proceeds much more smoothly than in an isotropic sample because all the crystalline domains are already oriented roughly along the axis of the fiber. Such orientation helps the growth of the domains, and smaller ones can more easily join up with each other. It explains the higher degree of crystallization of the fiber and, therefore, its better mechanical properties.

3.5 — Polymeric Liquid Crystals and Superstrong Fibers

We have now described how chemical fibers are produced. In many respects, such man-made fibers are no worse than natural ones. They are often used these days in the textile industry. Nevertheless, typical values of the breaking strength or the Young's modulus of polymeric fibers are one or two orders of magnitude lower than those of steel. So the question arises: Could we use the same physical idea as we have just described, but take it further to try to remove the difference? Is it possible to produce polymeric fibers of nearly the same strength as steel? Of course, the problem of creating such superstrong fibers is of great importance. There are many applications where light but strong materials are needed.

We can indeed make a superstrong fiber, even stronger than steel, from a polymer, but the polymer first must be converted into a special liquid-crystalline state which is really a variety of the viscous state. If you think of a viscous polymer as a "polymeric liquid," then a liquid-crystalline polymer can be regarded as an "anisotropic polymeric liquid." The anisotropy occurs spontaneously, with no help from the outside (such as orientating fields, mechanical stresses, or whatever).

Let's look at the simplest example (Figure 3.6) to see how this spontaneous orientation may appear. Just throw a bunch of randomly oriented matches onto a surface (Figure 3.6a). Now start reducing the area covered by the matches, but make sure that they remain oriented in the same random way. We gradually come to the situation in Figure 3.6b. At this stage it becomes impossible to decrease the area any further while retaining the orientational disorder. Does this mean that

a

b

c

FIGURE 3.6
Experimenting with matches on a surface: (a) random orientation at low concentration; (b) maximum concentration at which random orientation is still possible; and (c) orientational order at high concentration.

we have already reached a close-packed arrangement of the matches? Certainly not. They can be crammed into a much smaller square (Figure 3.6c), but their orientations would no longer be random; all the matches would be facing in the same direction. Hence, we conclude that the system of matches can be confined to a smaller area than in Figure 3.6b, but the system would then be anisotropic.

Now let us imagine that, instead of matches, we have a system (a solution) of molecules with an elongated shape. What will happen if we gradually raise the concentration of the solution? At lower concentrations we shall observe the pattern shown in Figure 3.6a—the distribution of the molecules' orientations will be isotropic. Then, as the concentration grows, we shall eventually reach the threshold regime as in Figure 3.6b. Obviously, at higher concentrations the solution can only be anisotropic. This anisotropy occurs for no external reason, but spontaneously, just because a dense enough system of elongated particles cannot possibly be arranged in any isotropic way.

This is exactly what a liquid-crystalline state is—an anisotropic state that spontaneously develops in a solution of elongated molecules at higher concentrations. Starting from some degree of asymmetry of the molecules (i.e., the ratio of length to diameter), the liquid-crystalline state can appear in a melt as well.

The name "liquid crystal" reflects the duality of such materials; according to their properties they could be placed somewhere in between ordinary liquids and crystalline solids. Like liquids, liquid crystals lack long-range order in the positions of their molecules; most liquid crystals are indeed fluid. At the same time, just like solid crystals, liquid crystals are anisotropic because their molecules are oriented in an anisotropic way.

It is clear from Figure 3.6 that the liquid-crystalline state should be more typical for substances whose molecules have an elongated shape. Moreover, the greater the asymmetry of the molecules, the lower the critical concentration of the solution at which the molecules start to align spontaneously. This suggests that solutions of stiff polymer chains should become liquid crystals rather easily, in quite a broad range of concentrations. Indeed, on larger scales such molecules are tangled up into coils. Therefore their asymmetry is determined by the ratio of the longest chain segment ℓ that can still be regarded as approximately straight (i.e., the Kuhn segment; see Section 5.5) to the characteristic diameter d of the chain. For rodlike polymer chains the ratio ℓ/d can be rather high; in the case of aromatic polyamides, for instance, it can even reach a few hundred. This means that even when the volume fraction of an aromatic polyamide is only a few percent, such a solution should still be liquid-crystalline. In other words, its molecular chains should be aligned predominantly along the axis of spontaneous orientation.

Now let us go back to the problem of how to produce superstrong fibers. A natural strategy is to use the inherent anisotropy of a liquid-crystalline solution and to form the fiber directly from this solution. If we do this, we shall end up with a highly oriented fiber immediately after the molding, extra stretching, and expulsion of the solvent. The orientational order will be much higher than normally achieved by orientational stretching alone. In practice, this technique allows liquid-crystalline solutions of aromatic polyamides to be converted into really amazing fibers whose strength and Young's modulus are of the same order of magnitude as those of steel. Such fibers were first created about 20 years ago, simultaneously both in Russia and the United States. Now they are widely used in various areas of industry.

3.6 — Polymer Solutions

So far when talking about various states of polymers, we have usually meant substances consisting purely of polymer molecules, although when discussing polymer fibers, we did mention that they are often formed from solutions of polymers.

Polymer solutions are, obviously, liquid mixtures of long polymer chains and small, light solvent molecules. They play a very important role in polymer physics; this is why it makes sense to give a brief description of them here. We shall discuss two qualitatively different uniform states of polymer solutions.

These states are illustrated in Figure 3.7; polymer chains are shown with solid lines, and small molecules of a solvent are not depicted at all. Figure 3.7a corresponds to a dilute polymer solution; macromolecules are separated by large distances and hardly interact with each other at all. The properties of such solutions are governed merely by the properties of the individual macromolecules. For instance, from light scattering or viscosity measurements, we can judge the shape and size of the polymer coils. So a polymer solution is, in a way, the most basic polymer system, because as we study it we actually learn about the properties of the individual macromolecules. In this sense, it is similar to a low-density gas of ordinary small molecules. Commonly, in more complex polymer systems the chains are highly entangled and strongly interact with each other; therefore, it is much harder to discern the contributions of individual macromolecules. To find out about the individual chains in this case, we would have to look at data for dilute solutions.

With increasing concentration, the polymer coils sooner or later start to overlap; then we eventually get to the picture of densely entangled coils shown in

FIGURE 3.7
(*a*) Dilute polymer solution.
(*b*) Intermediate regime between a dilute and a semidilute solution.
(*c*) Semidilute solution.
(*d*) Concentrated solution.
(*e*) Liquid-crystalline solution.

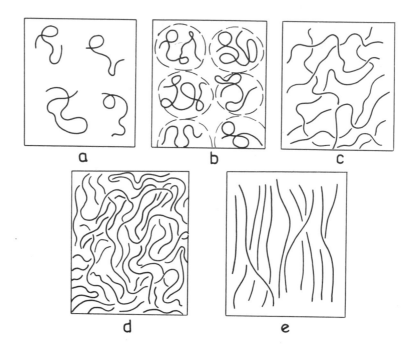

Figure 3.7*c*. Obviously, the intermediate regime between Figures 3.7*a* and 3.7*c* will be when the coils do not yet overlap, but just touch each other (as in Figure 3.7*b*). This means that the critical concentration $c^\star$ corresponding to the intermediate regime is the same order of magnitude as the concentration of monomers in each coil. It is useful to know how to calculate this value, or at least to estimate it. It would help us to understand which concentration regimes are realistic or typical for a polymer solution under different conditions. In Section 5.6 we shall come back to this question and find an approximation for $c^\star$. See also the movie on the CD ROM referred to in Section 5.6.

3.7 — Polymer Blends and Block-Copolymers

Polymer chains can, of course, mix not only with solvent molecules but also with other polymers. This is how polymer blends are formed. However, there are not that many pairs of different polymers that can blend in any proportion. A mixture of two polymers $\mathcal{A}$ and $\mathcal{B}$ will tend to separate into nearly pure phases of $\mathcal{A}$ and $\mathcal{B}$. This happens even if the repulsion between the monomers of $\mathcal{A}$ and $\mathcal{B}$ is so weak that they would be able to mix if they were not linked into a chain.

An interesting thing occurs when immiscible polymers $\mathcal{A}$ and $\mathcal{B}$ form one chain (Figure 3.8*a*). This is what we call a block-copolymer (see Section 2.5).

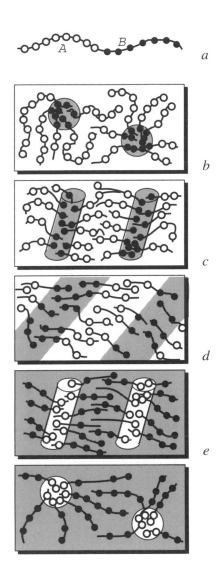

FIGURE 3.8
(*a*) A block-copolymer chain, consisting of two blocks, $\mathcal{A}$ (white) and $\mathcal{B}$ (black), (*b*)–(*f*). Possible types of microstructure in block-copolymer melts. Gray refers to the regions where black monomers dominate.

In this case the blocks $\mathcal{A}$ and $\mathcal{B}$ try to separate from each other. However, a macroscopic phase separation is impossible because the $\mathcal{A}$ and $\mathcal{B}$ blocks are tightly linked to each other within the chains. As a result, we get a pattern of microdomains that contain mainly $\mathcal{A}$ blocks or $\mathcal{B}$ blocks, separated by fairly thin interphase regions (Figure 3.8). This effect is known as microphase separation in block-copolymers, and the structure that emerges is called a microdomain structure.

Depending on the relative lengths of the $\mathcal{A}$ and $\mathcal{B}$ blocks, microdomains can take different shapes. In particular, they can look like little spheres (micelles), fitted onto a regular three-dimensional lattice (Figure 3.8*b*). This happens if the $\mathcal{A}$ blocks are much shorter than $\mathcal{B}$ ones. Increasing the length of the $\mathcal{A}$ blocks changes this picture, and we end up with cylindrical $\mathcal{A}$ domains in a "sea" of $\mathcal{B}$ units (Figure 3.8*c*). When the two blocks have roughly the same lengths, there are alternating $\mathcal{A}$ and $\mathcal{B}$ layers (Figure 3.8*d*). Finally, for shorter $\mathcal{B}$ blocks, cylindrical (Figure 3.8*e*) and spherical (Figure 3.8*f*) $\mathcal{B}$ "islands" appear in the "sea" of $\mathcal{A}$ units. Thus, we have a tool for controlling the microstructure of a block-copolymer melt. All we need to do is vary the length ratio between the $\mathcal{A}$ and $\mathcal{B}$ blocks.

3.8 Ionomers and Associating Polymers

A rich variety of microstructures and inhomogeneities is very typical of polymers. Fully uniform states are the exception rather than the rule. Now we shall explore yet another peculiarity of polymer structure, by looking at the so-called ionomers.

We have already discussed polyelectrolytes in Section 2.5. They are formed when small ions, called counterions, break off from the chain. They leave behind monomer units of the opposite charge. If a counter-ion escapes and sets out on a "journey" on its own, the whole chain acquires an electrical charge and becomes a polyelectrolyte (Figure 3.9*a*). However, this is not the only scenario. Thermal motion may not be large enough for the counterion to tear itself away from the ionized monomer. Instead, the two form an "ion pair." The counterion stays in the vicinity of the charged monomer (at an average distance *a*), the two charges making a dipole (Figure 3.9*b*). If all the counterions tend to stay in such pairs, the chain is called an ionomer.

Can we tell exactly when each of these two cases, a polyelectrolyte (Figure 3.9*a*) and an ionomer (Figure 3.9*b*), would occur? Assume that the charges of the

FIGURE 3.9
(*a*) A polyelectrolyte: all the counterions are free and not attached to the polymer chain. (*b*) An ionomer: counterions are "condensed" on the charges of the chain and form ion pairs.

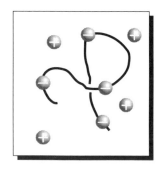 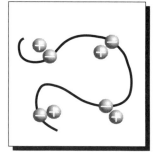

a   b

dissociated monomer and the counterion are the same in magnitude and equal to the electronic charge e. Suppose also that the dielectric constant of the medium is ε. Then the energy of the Coulomb interaction of the ions in a pair is $e^2/\varepsilon a$. If this energy is much less than the characteristic energy of thermal motion kT (where k is Boltzmann's constant and T is the absolute temperature),

$$\frac{e^2}{\varepsilon akT} < 1, \tag{3.2}$$

then counterions break off the chain. Thus we get the polyelectrolytic regime. Otherwise, if

$$\frac{e^2}{\varepsilon akT} > 1, \tag{3.3}$$

then the thermal motion cannot break up the ion pairs and the chain is an ionomer.

We have already said that proteins, DNA, and polyacrylic and polymethacrylic acids are all polyelectrolytes when dissolved in water (see Section 2.5). We stress that the solvent should be water. Why is this so essential? The dielectric constant of water is extremely high ($\varepsilon \approx 81$). Therefore, the ratio $e^2/(\varepsilon akT)$ is relatively small, and inequality (3.2) holds. However, if you use other solvents, with much smaller values of ε (usually between 2 and 20), then inequality (3.3) holds instead, and the polymer is an ionomer rather than a polyelectrolyte.

Ion-containing polymer chains in a melt (in the absence of a solvent) are also in the ionomer regime. This is because the dielectric constant of a pure polymer tends to be rather low. What is the structure of such a melt? Ionomer chains contain some (usually small) proportion of monomers in the form of ion pairs (see Figure 3.9b). They interact strongly with each other, since they are electric dipoles. The other monomers have no electrical charge. Dipoles always arrange themselves in such a way that the interaction between them is attractive (see Figure 3.10). This is why ion pairs are strongly attracted to each other. However, they are part of a polymer chain, so they cannot be separated into a distinct phase.

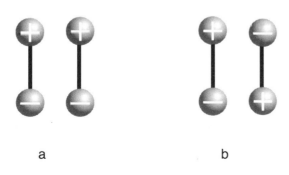

a   b

FIGURE 3.10
Dipoles (ion pairs) are free to choose any orientation with respect to each other. The one they prefer is (b), since it gives the lower energy. It corresponds to attraction.

As a result, small islands emerge in a sea of neutral monomers (Figure 3.11). Such aggregates are called ionomer multiplets.

Compare Figure 3.11 and Figure 3.8b. You may notice that the multiplets in a polymer melt resemble the spherical domains in a block-copolymer melt that form when one of the blocks is much longer than the other. This similarity is not surprising. The structure in Figure 3.11 can be obtained from Figure 3.8b by letting the shorter block tend to one monomer and increasing the attractive force between the monomers. As a matter of fact, this is quite a common pattern in polymer structure when small spherical multiplets ("associates") are formed by strongly attracting monomers. Such polymers are called associating.

Associating polymers have many practical uses. Let us give the simplest example. Suppose we wanted to increase the viscosity of a liquid substantially. Can we do it by adding just a little bit of some other substance? If yes, then how do we choose the substance? As you might have guessed, an associating polymer consisting of two different types of monomers (Figure 3.12) can play this role. Let's see how it works. The greater part of the monomers easily dissolve in the liquid. However, a small fraction of the monomers try to avoid the solvent. Any contact with the liquid molecules is extremely energetically unfavorable for them. Such strongly associating monomers join together to form aggregates (multiplets). The structure of the resulting solution is shown in Figure 3.12. It looks like a polymer network (a gel). The special thing about it is that the role of covalent bonds is played by associates of strongly interacting monomers. This is not a real network (like the ones described in Section 2.5). The associates can dissociate from time to time, as well as appear in a new place. Therefore, if we apply an electrical voltage across the liquid, it will start flowing, together with the associating polymer

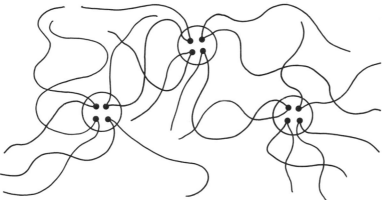

FIGURE 3.12
The structure of a solution of an associating polymer. Strongly interacting ion pairs are depicted with big dots. Associates are circled. In contrast to Figure 3.11, the space between associates is mainly taken up by the solvent molecules. (The volume fraction occupied by monomer units of the chains is rather small.)

that is dissolved in it. Nevertheless, the viscosity of the liquid is substantially increased because dissociation of such associates is relatively rare. Even a very small amount of an associating polymer is enough to increase the viscosity of the liquid substantially; Figure 3.12 illustrates this. As you can see, most of the volume is occupied by the liquid itself (it takes up all the space between the associates).

With this we conclude our brief review of the kinds of states in which the simplest polymers can exist. Of course, we did not cover the whole variety of polymer systems. A few more examples will be found further on in this book, as well as in some other popular books listed in the Suggested Readings.

4

Polymers in Nature

In a bath, in a tub, in a
shower,
In a stream, in a brook,
in the sea,
Here and there, and
everywhere—
Glory to water forever
be!
K. Chukovskiy, *Wash-
into-holes* (Russian
children's poem)

A great many fascinating biological objects made up of polymers. For example, the shell of a tortoise and the stiff back of a beetle are "built" from a polymer called chitin whose chains are held together by proteins (which are polymers too). Then there are viruses—little boxes made from protein chains, with a nucleic acid chain inside each. There are far too many more examples to tell of them all! We shall therefore have to stick to three, ones which we, as physicists, believe are the most interesting and fundamental.

However, before we start our story, there is one more thing to say: The main biopolymers function in the medium of water. A human body consists of 60% water by mass; some animals carry around even more water in their bodies. Water reservoirs are a source of life (as we shall discuss in more detail in Chapter 11). Therefore, it might be helpful to learn a bit about the molecular structure of water, before we plunge into the discussion of biopolymers.

4.1 — A Few Words about Water and the Love or Fear of It

A molecule of water, H_2O, is triangular in shape (Figure 4.1). The electron cloud tends to be shifted away from the hydrogen nuclei toward the oxygen nucleus by, on average, $0.2 \text{ Å} = 2 \cdot 10^{-11}$ m.

FIGURE 4.1
A molecule of water.

As a result, the positive charge of the hydrogen nuclei is not quite compensated. Similarly, there is an uncompensated negative charge around the oxygen nucleus. This peculiarity of the structure may not seem of great significance at first sight. However, it is the real cause of all the special properties of water that make it play such an important role in living organisms. What are these properties?

First, a water molecule has a considerable dipole moment, $p = 0.6 \cdot 10^{-29}$ Cm, so the molecule is polar (this is what we call substances whose molecules have a nonzero dipole moment). This means that in an external electric field water molecules can be regarded as little "dipoles," each carrying two charges, $+e$ and $-e$, separated by a distance a (e is the charge of electron, $e = 1.6 \cdot 10^{-19}$ C); then $p = ea$. Given the value of p mentioned above, we can calculate $a = p/e = 0.4$ Å $= 4 \cdot 10^{-11}$ m. Such little dipoles have no difficulty in becoming aligned in an external electric field; this explains why the dielectric permeability of water is much higher than for all other common liquids: $\varepsilon \approx 81$.

In Section 3.8 we decided that such a high value of dielectric constant means that many monomers dissociate in water solutions. In other words, the corresponding polymers are polyelectrolytes. In particular, the polyelectrolytic nature of the main biopolymers, DNA and proteins, is crucial for their biological functioning.

Second, water molecules appear to be able to form so-called hydrogen bonds between each other. A hydrogen bond is a kind of saturable, attractive interaction between a couple of atoms, say O, C, N, and so on. One of the two atoms should be joined to a hydrogen atom by a covalent bond. For instance:

$$O - H \cdots O,$$

where the dots mark the hydrogen bond and the solid line the covalent bond. Roughly speaking, the attraction occurs because the hydrogen atom's electron is shifted toward the oxygen atom along the covalent bond. As a result, there is some extra positive charge near the H nucleus as well as some extra negative charge around the O nucleus. Thus an H nucleus can be attracted to an O nucleus of another molecule, linking the two molecules together. The binding energy of a hydrogen bond is of order 0.1 eV $= 1.6 \cdot 10^{-20}$ J; this is one or two orders of magnitude smaller than a covalent bond's energy (which is about $1 - 10$ eV), but somewhat larger than the thermal energy at room temperature (300 K): $kT \sim 0.03$ eV where $k \approx 1.38 \cdot 10^{-23}$ J/K is Boltzmann's constant. Therefore, the molecular structure of water at any instant just looks like a three-dimensional network of hydrogen bonds (Plate 3). However, in contrast to a gel, this network gets torn apart and stuck together in a new manner over and over again, due to the thermal motion.

The network of hydrogen bonds is a key concept clarifying many properties of water, such as water's high heat capacity. Indeed, to increase the temperature of water you have to expend a fair bit of energy to break the hydrogen bonds.

What we have said about water also explains its special features as a solvent. Nonpolar substances (i.e., substances whose molecules have no dipole moment, such as the simplest organic compounds—fats and oils) are barely soluble in water, whereas the solubility of polar substances is normally much greater. This can be explained in the following way. If a polar molecule is placed in water it experiences a strong attraction to the water molecules. This is due to the interaction between the little dipoles, which try to line up parallel to each other. For a low molecular weight molecule, the energy of such attraction is usually around 0.1 eV and is quite often enough to provide significant solubility. In contrast, if there is a nonpolar molecule in the water, there will be no attraction. In fact, just the opposite will occur—the water's molecular structure will be distorted as some hydrogen bonds will be broken. Obviously, this is not energetically favorable, and so the water molecules will try to "push" the alien molecule out. Such molecules have practically zero solubility.

Polar and nonpolar substances are also known respectively as hydrophilic and hydrophobic. These names start making sense when translated from Greek: *hydor* of course means water, *philos* means love or attraction, and *phobos* means fear. The concept of hydrophilic and hydrophobic behavior is very important in present-day molecular biology.

4.2 — Head-and-Tail Molecules

Add some hydrophilic substance to a glass of water, and it will merely mix with the water, just like sugar. In other words, it will be dissolved. On the other hand, a hydrophobic substance cannot be dissolved; it will separate out from the water, just like oil. However, there is a more complex "amphiphilic" kind of molecule; each molecule contains both a hydrophilic and a hydrophobic part. What happens to them in water?

FIGURE 4.2
A schematic diagram of
a typical amphiphilic
molecule.

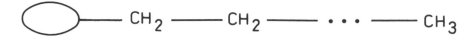

Each of us have witnessed many times the interaction between amphiphilic substances and water, since even ordinary soap consists of amphiphilic molecules. (How could we avoid mentioning soap having chosen an epigraph from the book *Wash-into-holes*?) Besides, amphiphilic molecules are often encountered in biological systems. Most often such molecules consist of a polar atomic group, the "head" (Figure 4.2), and a hydrophobic "tail," which is attached to the head. The tail is a carbohydrate chain $(-CH_2-)_n$ of medium length; normally n varies in the range from 5 to 20. The whole molecule looks very much like a tadpole. Strictly speaking, "tadpole" molecules like the ones in Figure 4.2 are not quite polymers since the number n is not high enough. Nevertheless, it is just due to the presence of the tail that such molecules have some rather special and interesting properties.

    Watch the movie called *"Soap"* on the CD ROM. We show there why soap can wash things.

So what will happen if you try to dissolve molecules like those in Figure 4.2 in water? A straightforward guess is that while there are not too many of them they will stay on the surface, immersing their heads in the water and sticking their tails out of it (Figure 4.3a). By the way, this makes it clear how soap actually works. Oil, fat, and other nonpolar organic compounds cannot be easily washed off by water because they just do not dissolve in it. However, there is a great difference as soon as amphiphilic molecules of soap come along. Their hydrophobic tails will cling to the oil because it is energetically favorable for hydrophobic particles in the water to come together. For them, getting together is simply a way of defending each other from being too close to the water. As a result, the water molecules form

FIGURE 4.3
The behavior of
amphiphilic molecules
in water.

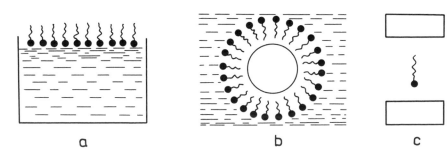

a kind of "coating" around the drops of oil (Figure 4.3*b*). The whole surface of such "coated" particles consists of the hydrophilic heads of the soap molecules; therefore they are soluble and easily washable by water. Thus, in a sense, the amphiphilic soap molecules stick the oil to the water (Figure 4.3*c*).

The next question to ask is: What if there are too many "tadpoles" and they cannot all be accommodated on the water's surface? From our own experience with soap we know that the surface area becomes considerably larger if a lot of bubbles or foam appear. (Such substances are even referred to as surface-active). However, if there is no way to further increase the surface area, then the approximate scenario shown in Figure 4.4 gradually occurs with the increase in the "tadpole" concentration.

The first stage is that the "tadpoles" get together to form spherical particles, called micelles. Each micelle's outer surface is made up of hydrophilic heads and is in direct contact with water, whereas the hydrophobic tails are hidden inside (Figure 4.4*a*). Obviously, such micelles dissolve easily in water (because to water they seem purely hydrophilic). On the other hand, they behave like highly stable, almost indestructible units.

**The formation of micelles is shown in the special movie called
"*Micelles*" on the CD ROM. Red particles are hydrophobic, so they
tend to aggregate. On the other hand, green particles are
hydrophilic, so they tend to be exposed to surrounding water
(which is not seen). The conflict of these two tendencies leads to
the formation of micelles, where hydrophobic particles are hidden
from the water.**

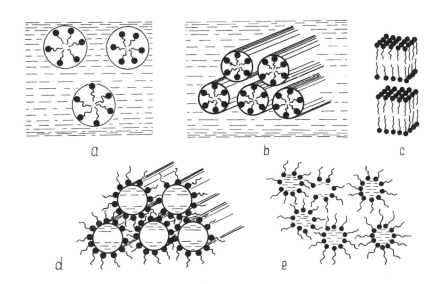

FIGURE 4.4
**The change in structure
of solutions of
amphiphilic molecules
with increasing
concentration.**

If there are even more amphiphilic molecules in the solution, the spherical micelles start to feel rather cramped. The "tadpoles" reorganize to form a system of parallel cylindrical micelles (Figure 4.4b). Now imagine that the number of "tadpoles" continues to grow. You can regard it as just having less water—no longer enough to fill in the gaps between the cylindrical micelles. The amphiphilic molecules are forced to rearrange themselves once again, this time forming parallel layers known as lamellae (Figure 4.4 c). If there is even less water, then inverted cylindrical micelles (Figure 4.4d) and, later, inverted spherical ones (Figure 4.4e) gradually develop. We end up with lots of different and very beautiful structures!

Substances with such structures have quite unusual properties. They are fluid, but in the cases shown in Figures 4.4b and 4.4d they tend to flow differently parallel and perpendicular to the cylinders. Meanwhile, for the lamellar structure of Figure 4.4c there is only one possible direction of flow—the layers can only slide parallel to each other. Certainly, light too propagates differently along and across the cylindrical micelles or the lamellae; hence birefringence is typical.

To tell the truth, you cannot usually get all five successive stages (Figures 4.4a–4.4e) with the same substance. Either the tails are too thick to form spherical or even cylindrical micelles, or, on the contrary, they may be too thin to construct inverted micelles. Normally one substance can exhibit only two or three of the structures in Figure 4.4.

Let's compare Figures 4.4 and 3.8. You can easily spot the similarity between the structures that "tadpoles" form in water and those appearing in block-copolymer melts. This is not a coincidence. Indeed, block-copolymers are also amphiphilic molecules, just like the "tadpoles." The only difference is that, instead of a tail and a head, they simply have two tails connected with each other.

The structures like the ones in Figure 4.4 are often used in a special kind of polymerization process called emulsion polymerization. In Section 2.7 we described how polymerization occurs. We discussed that the chain often "breaks" because its two growing ends come together. If we knew how to make the ends less likely to encounter each other, we would be able to produce longer polymers. One of the solutions is to carry out polymerization in a system such as that shown in Figure 4.4a. Assume that both the initiator and the monomers are insoluble in water, but that they do not mind the tadpoles' hydrophobic tails. Then, if you dissolve them both in the system shown in Figure 4.4 a, they will be mostly absorbed by hydrophobic micelles. Polymerization will start in the micelles where the initiator molecules ended up. You can adjust the concentation so that there is no more than one molecule of the initiator in most micelles. Then the chain will be safe from breaking. The polymerization will go on until there are no monomers left in the micelle. Thus, we create longer polymers than if we use one of the usual methods. Moreover, the chain tends to grow faster, because monomers are trapped in a special "microreactor," a micelle. Since this microreactor has a microscopic size, it helps to solve another problem—how to take away the heat that is given out during the reaction.

This is all fairly interesting, but you may start wondering what it has to do with biology. Here is the answer: Molecules of phospholipids have the shape of a "tadpole," although normally with two, or sometimes even three, tails. They are the chief constituent of membranes, special walls that separate biological cells from the outside world and divide the cells into sections. The considerable thickness of the double tails prevents phospholipids from clumping into micelles, hence they form into layered walls.

Phospholipids can even be used as a material to make a model of a real cell. All you have to do is to take a suspension of phospholipids and to give it a good "shake" with an ultrasound signal of the appropriate wavelength. This forms liposomes, which are comparable in structure to that shown in Figure 4.5. Liposomes are used, for example, to study how different drugs may penetrate into a cell through the cell membrane.

However, the phospholipid layer is not the only part of a membrane. There are also some proteins "floating" in the lipid medium (Figure 4.6). Moreover, the membrane (and thus the whole cell) is held in shape by the so-called cytoskeleton, which consists of proteins and polysaccharides (which are polymers too!) The strange name comes from the Greek for cell, *cytos*.

The study of cell membranes is one of the most rapidly developing branches of modern biology; it even has its own name, membranology. There is a great variety of interesting phenomena in this area, and quite a few of them are related to polymers. Unfortunately, there is no way to describe them all in this book, but

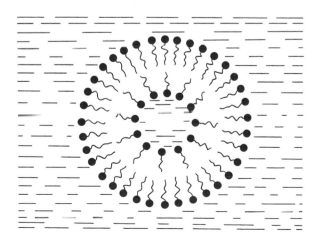

FIGURE 4.5
A liposome.

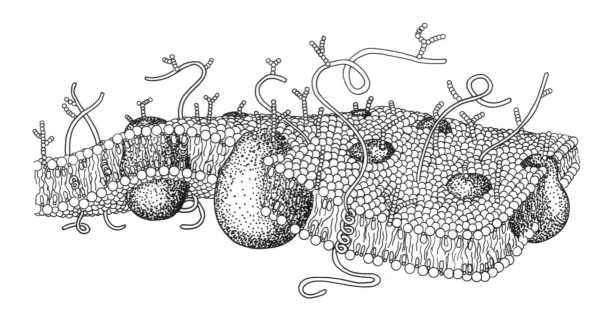

FIGURE 4.6
**The structure of a
biological membrane.**

we cannot help giving one particular example. In principle, the "fatty" layer of a membrane can exist in two different states. One of the states is nearly solid, with the hydrophobic "fatty" tails lying parallel to each other. In contrast, the other state is liquid, and the tails are randomly entwined. There is a possibility of a phase transition between the two states. (To be precise, the cytoskeleton is actually involved in this transition too.) At different stages of a normal cell's life cycle, its membrane appears in the two different states, sometimes changing from one to the other. (A cancer tumor cell, however, which is prone to uncontrollable division, is incapable of such transitions, so its membrane remains liquid all the time. This may turn out to be important in understanding cancer, although as yet we do not know why.)

Thus, some rather complicated structures can be built from phospholipid molecules in nature. Later we shall see even more interesting "architecture" when discussing proteins and nucleic acids. However, we first need to explain what made us mention architecture.

4.3 — Molecular Biology and Molecular Architecture

In books on the history of architecture you may come across an interesting theory. People who are keen on scientific explanations may find it quite attractive. In our own words, it is the following.

How could you work out, if you wanted to, what kind of architectural style was typical of some period in history? It turns out that you do not really have to study the aesthetic views of that epoch. All you need is just some knowledge of the mechanical properties, the elasticity and strength, of the building materials used at that time.

To make this clearer, we give some examples. A structure made of huge unattached stone blocks appears tremendously strong in compression, but very weak in shear; bending (torsion) may only be felt by some individual blocks. The Egyptian pyramids in which the pharaohs were buried are an extreme example of this method of construction. They have the most pointed tops that can be created using material that cannot withstand shear forces.

On the other hand, for a building without a pointed top, columns (which work in compression) are used to support continuous beams (working in bend). This is exactly how the glorious Parthenon in Athens was constructed. Layers of bricks or a system of small stone blocks fixed to each other would still be very strong in compression, and satisfactory in shear. However, they would not work in tension or in bend at all. Vaults of a ceiling made in such a way have to arch upward. This kind of design can be seen in the Hall of Facets in the Moscow Kremlin and in white stone churches all over Russia. It also occurs in the Gothic cathedrals of Western Europe. In fact, numerous little towers and buttresses are needed to avoid parts of the walls being in tension, by introducing more compression.

Wooden logs are a different example—they are strong in compression perpendicular to their length and in tension along it. Hence, apparently, were built the wooden temples in the North of Russia. And finally, reinforced concrete can work beautifully in all types of deformation, which explains the giant vertical and horizontal surfaces in modern constructions, such as the United Nations building in New York City and the former Council for Economical Cooperation building in Moscow. Of course, whatever style and material, there may be all sorts of architecture—some buildings show no spark of talent, whereas some others are real works of genius. But that is a very separate discussion indeed!

Back to biology: If we think of a chemical substance as a kind of architectural construction and the molecules as the building material, we shall get quite a similar situation. In the previous chapter we looked at how properties of various polymer chains are determined, in the end, by the chain structure of the individual molecules. However, no architect in his worst nightmare would ever dream of becoming a polymer technologist. The trouble is that a polymer scientist is unable to pile up the molecules, one by one, in specially determined places as if they were bricks or logs. Instead, indirect methods must be used, such as heating and cooling or dilution and sedimentation, to encourage the molecules to arrange themselves in a way at least vaguely similar to what is desired. (Just imagine an architect trying to build something sensible out of a pile of bricks by merely shaking it, or by floating the bricks in water and pouring them out!) This is why the order of molecular segments and the geometrical structure of synthetic materials (particularly polymers) always has many faults and can never be perfect. (For example, if entwining polymer chains in a fiber could be arranged to the perfection of a tidy little girl's hair plait, then we would have fibers of amazing strength—as much as an order of magnitude stronger than that of the best known ones made from a liquid crystalline solution.) However, this is only the case with artificial materials.[1]

The situation in biology is totally different. Molecular biology does have something in common with molecular architecture. First, a number of molecules in biological systems are such that they manage to naturally keep very high order among themselves (we have seen it with lipids). Second, there are some special systems in a living cell that are capable of arranging molecules in a given structure; an example is a ribosome, which is responsible for protein synthesis. This is why a peculiar novel language comes into use when a physicist starts talking about proteins and nucleic acids—rather unusual and different from all the rest.

4.4 – Molecular Machines: Proteins, RNA, and DNA

The special thing about biological macromolecules (proteins, RNA, and DNA) is that they have biological functions to fulfill. You could say that RNA and DNA are not only molecules of a particular substance, but also a device or a machine to do particular operations. In this sense it is more straightforward to talk about such polymers in the language used in cybernetics to describe robots.

In particular, as we have already said, a strictly fixed sequence of different monomer units in a chain of a biopolymer can naturally be compared to a text written with the appropriate molecular "alphabet." Since such a sequence chemically determines the individuality of, say, a protein, then in the spirit of this analogy we can say that a "protein text" lists or codes the function of the protein. The sequence of monomer units in DNA, as everyone knows, contains genetic information, and it codes the "texts" of proteins by means of the so-called genetic code. This is the cybernetic terminology commonly used in molecular biology.

---

[1] Look now at Figure 10.1*b* and Plate 14 for yet another parallel between molecular and architectural patterns.

This cybernetic analogy is beautiful and comprehensive, but it does not tell us anything about the way in which the processes actually occur. Why can one protein with a given piece of text detect photons of light in the retina of an eye, and another with a different piece of text cause the physical effort of a muscle, whereas a third controls the immune system, and a fourth...? (One cannot easily count all the functions of proteins: catalysis of strictly specific reactions, including the biosynthesis of proteins and DNA; strictly specific transportation of molecules through membranes; tangling and disentangling of knots on DNA, etc. As a matter of fact, all the processes in a cell are carried out by proteins.) So how is the DNA text read, and how, according to the instructions that it contains, is a protein built? Certainly all these and similar questions are connected with the physics of biopolymers. We already know a lot, but there is still far to go to reach a complete understanding. Perhaps you, however young or old, may someday take part in these studies. For now, we shall just describe in brief what is already known.

4.5 — The Chemical Structure of Proteins, DNA, and RNA

First, a few words about the chemistry of the subject. Monomer units of a protein chain are residues of the so-called amino acids and have a structure of the sort $— CO — CHR — NII —$. Here R stands for a radical, which can be of 20 possible types. In the simplest case it is just a hydrogen atom, and the corresponding amino acid residue is called glycine (Gly). For the remaining 19 amino acids the radical R has a more complex structure, such as $— CH_3$ (alanine, Ala), $— CH_2OH$ (serine, Ser), $— CH_2 — CH_2 — S — CH_3$ (methionine, Met), $— CH_2 — CO — NH_2$ (asparagine, Asn), $— CH_2 — COO^-$ (aspartic acid, Asp), and $—(CH_2)_4 — N^+H_3$ (lysine, Lys). The latter two examples show that protein chains may contain monomer units that carry a positive or negative electrical charge. (For example, a charged unit may be created when a small molecule dissociates, losing a low molecular weight ion.) The sequence of amino acid residues in the chain is different for different proteins; you can regard this sequence as a kind of "text" written in a 20-character protein "alphabet." The number of monomer units N in each protein molecule usually varies from a few tens up to a few hundreds.

The spatial structure of a couple of protein chain units is sketched in Figure 4.7. The $— CO — NH —$ bond links together $— CHR —$ groups that are specific to each unit. It is called a peptide bond; that is why the whole protein molecule is often referred to as a peptide chain.

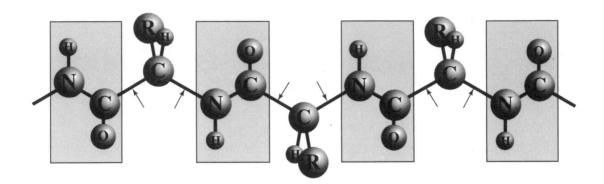

FIGURE 4.7
The chemical structure of a protein chain. In the picture, flat amide groups are enclosed in rectangles. They all contain —N—C— peptide bonds, which form the main chain backbone when the chain is finally complete. The arrows show the bonds about which the chain may rotate. (The corresponding angles of rotation are called Ramachandran's ϕ and ψ angles.) R symbolizes side groups of various amino acid residues.

The chemical structure of DNA strands is illustrated in Figure 4.8. Each strand is made up of alternating sugar (deoxyribose) and phosphate groups; due to the latter, the whole strand has a negative charge. A nitric base is attached to each sugar group. There are four possible bases: adenine (A), cytosine (C), guanine (G), and thymine (T). They are all shown in Figure 4.8. RNA strands have a similar structure, with a different type of sugar in the main chain and the base uracil (U) replacing thymine.

As for the three-dimensional structure, you probably know that in a living cell, DNA molecules consist of two strands (like the one in Figure 4.8) that form a double helix (see Figure 4.10c). It is essential that these two strands are mutually complementary. This means that, say, adenine in one of the strands always corresponds to thymine in the other, whereas guanine always corresponds to cytosine. Physically, the reason for this is that the nitric bases are located in the very core of the double helix, where only the pairs A–T and G–C can fit perfectly without distorting the shape of the double helix. Figure 4.9 explains why this is so. Hence, the second strand of the double helix contains no extra information, but merely helps to reproduce the information and to make multiple copies of it.

4.6 — Primary, Secondary, and Tertiary Structures of Biopolymers

Going into the physics of biopolymers, the first thing to understand is the hierarchy of their structure. As you must have guessed from the title of this section, there is a primary, a secondary, and a tertiary structure, and sometimes even a quaternary one.

The primary structure, as we have already mentioned, is the sequence of repeat units in the chain, that is, the "text." It is created during the biosynthesis

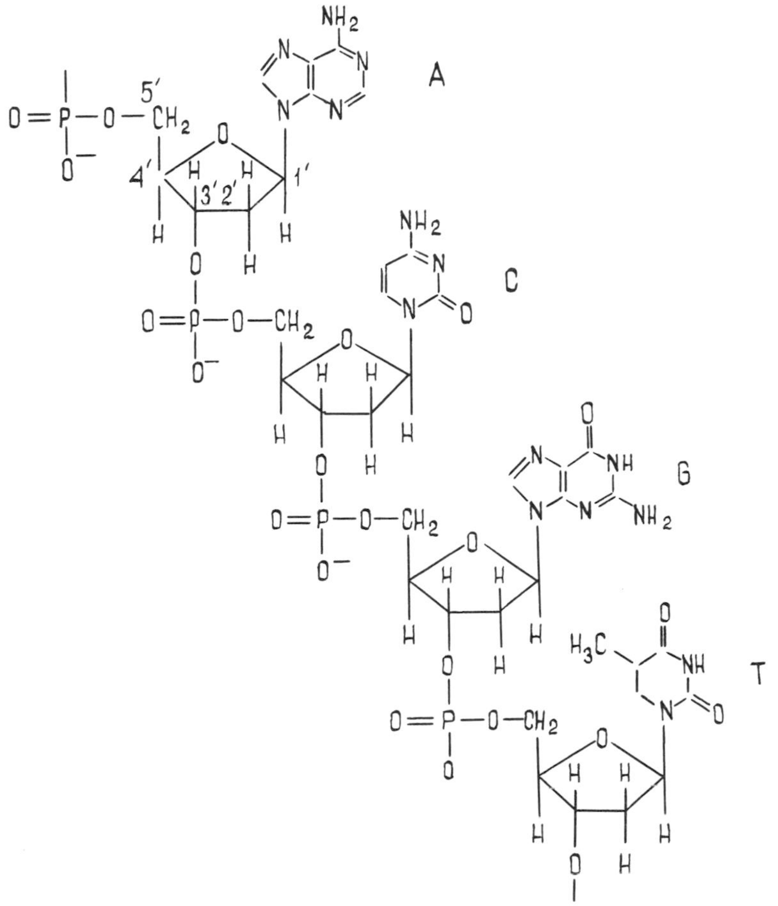

FIGURE 4.8
The chemical structure of a single strand of DNA.

of each molecule and is "memorized"; in other words, it is not distorted by the thermal motion unless the whole molecule is destroyed. Biochemists have learned how to find the primary structure of a protein molecule as well as that of DNA. To some extent, these experiments can be now carried out automatically, and there is a huge, constantly increasing international database of the corresponding data. A project has even been embarked on to determine the complete primary structure of a human's DNA molecule, that is, a human's genome (hence the name Human Genome Project)! Unfortunately, it is extremely difficult (and at present hardly possible) to carry out *in vitro* a chemical synthesis of long protein chains with a given primary structure.

Some hopes of tackling this arose in 1984 when the English scientist A. Fersht came up with a totally new approach. His idea was to use a natural system of biosynthesis that is at work in every living cell, a method is called protein

FIGURE 4.9
**Explanation of
the mutual
complementarity of the
two strands of the
DNA's double helix.**

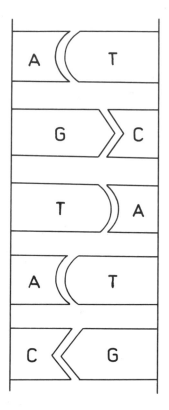

engineering. In principle, we could get a cell to produce protein molecules with any primary structure we wish.[2]

However, even if the technical side were no problem, the trouble is that no one really knows what to wish for. When the total number of units $N \sim 10^2$ and there are 20 "candidates" for each unit, we end up with approximately 20^{100} different possibilities. How can we choose some sensible ones out of such a tremendous number? Of course, one can merely "duplicate" some genuine proteins just as they exist in nature. However, this would be quite an extravagant pastime; it would be only worth trying if we managed to "improve" on nature. (It is a bit like a publisher who is unable to read: He is free to publish any book, but how would he choose an exciting and informative text that would become a best-seller?) As for biopolymers, it only becomes clear whether their texts are "exciting and informative" when the secondary and tertiary structures are formed.

----

[2] The technique of genetic engineering that had appeared some ten years before, is not quite as powerful—it enables one cell to make a protein molecule that is typical of another cell. For example, a bacterium can be used to synthesize human insulin.

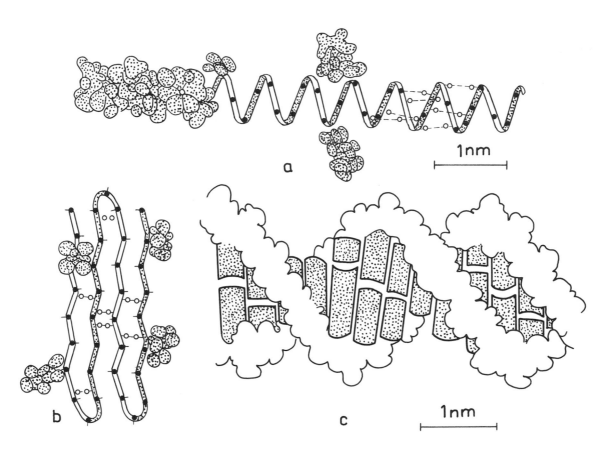

FIGURE 4.10
Secondary structures of biopolymers: (a) α-spiral of a protein, (b) β-structure of a protein, (c) DNA double helix. To make the image clearer, together with the main polypeptide chain, the positions of some side groups are also shown in (a) and (b).

Secondary and tertiary structure are the short-scale and long-scale order in the monomers' positions, respectively. The main secondary structures of proteins, the α and β structures, were discovered in the 1940's and1950s by the American chemist Linus Pauling (who later received the Nobel prize in chemistry for these studies; by the way, he also won the Nobel Peace prize). The normal secondary structure of DNA—the famous Watson and Crick double helix—was discovered in 1953 by Francis Crick and James Watson at the University of Cambridge in England. All the structures mentioned are sketched in Figure 4.10. They are made stable by the hydrogen bonds. In fact, the reason why the loops of the helix and the β-folds are formed is simply that this is the arrangement that achieves the maximum saturation of the hydrogen bonds.

The discovery of the double helix has won a firm and very fair reputation as one of the major achievements in the history of science. The striking beauty of the double-helix model of DNA is the way it explains, with brilliant ease, one of the real marvels of nature—the ability of all living things to reproduce themselves. Indeed, as soon as the two complementary strands move apart, they form something like a pair of "printing plates" or templates that are ready to make two identical copies. This is exactly how biological inheritance occurs at the molecular level.

There can be some unfavorable conditions when secondary structures do not develop in biopolymer chains. The physics of this question is of great interest. For instance, if the temperature is increased, or some low molecular weight substances are added to the solution, it can cause untwisting of the helices, called a helix-coil transition. This transition has this name because a spiral-shaped polymer chain is rather rigid, whereas a nonspiral one is a relatively flexible coil (Figure 4.11). Therefore the helix-coil transition is also called the melting of the helix. The analogy with the ordinary melting of solids is even more appropriate since the helix-coil transition occurs very rapidly with increasing temperature, over a narrow temperature range. It is accompanied by considerable absorption of heat, which is used to break the hydrogen bonds forming the helix. Meanwhile, during melting, not only are individual loops of the helix destroyed (this would be similar to the melting of individual cells of a crystalline lattice), but whole chunks of the helix collapse. This is a cooperative effect; the loss of one loop helps the next one to fall apart. However, in some ways the helix-coil transition is different from ordinary melting. The main difference is that spiral and nonspiral strands do not separate out (as, say, regions of ice and water do in a winter river), but they are mixed along the chain. In the language of theoretical physics, we can say that this is not a phase transition.

The theoretical interpretation of all this is rather interesting. Apparently, the helix-coil transition is indeed a real melting process, although not of a three-dimensional crystal but of a one-dimensional one. In the one-dimensional world,

FIGURE 4.11
DNA strands: (*a*) in a double helix, (*b*) in the molten state.

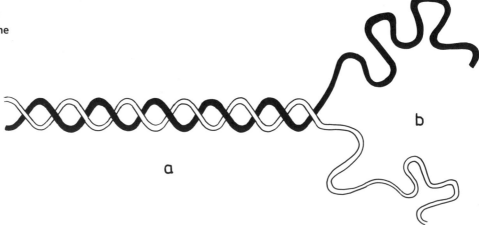

b

a

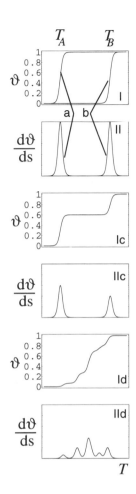

FIGURE 4.12
Melting curves (I, $\theta(s)$) and differential melting curves (II, $d\theta(s)/ds$) for different sequences: (*a*) poly-$\mathcal{A}$ homopolymer, (*b*) poly-$\mathcal{B}$ homopolymer, (*c*) block-copolymer with 60% A and 40% B, (*d*) heteropolymer with random alternation of the monomer units.

melting is a rather rapid process, but it does not lead to phase separation. This fact is known to physicists as Landau's theorem.

It is also interesting that a real heteropolymer with a nonuniform primary structure does not melt as sharply as a specially prepared homopolymer. Figure 4.12 explains why, by comparing polymers with different primary structures and showing their melting curves. These are dependences of the spiral coefficient θ (i.e., the fraction of spiral units in the chain) on inverse temperature T^{-1}. Two uniform homopolymers, say $\mathcal{A}-\mathcal{A}-\cdots-\mathcal{A}$ and $\mathcal{B}-\mathcal{B}-\cdots-\mathcal{B}$, both melt rather dramatically, but at different temperatures—see Figures 4.12*a* and 4.12*b*. (For example, the difference in melting point for DNA molecules that consist only of A–T pairs or only of G–C pairs is as big as 40°C). Obviously, a copolymer $\mathcal{A}-\mathcal{A}-\cdots-\mathcal{A}-\mathcal{B}-\mathcal{B}-\cdots-\mathcal{B}$ would melt in two stages (see Figure 4.12*c*). Hence it is not surprising that the melting of a real heteropolymer with a complex sequence of monomers $\mathcal{A}$ and $\mathcal{B}$ is a gradual process (Figure 4.12d).

These differences in behavior become even more obvious if we look at the so-called differential melting curves—dependences of the derivative $\partial\theta/\partial T$ (i.e., the tangent to the curves in Figure 4.12, I) on T. Differential melting curves normally look like a set of peaks (Figure 4.12, II): a peak at some temperature T_0 is associated with the melting of a particular strand of the helix—namely, the one whose primary structure is such that the ratio of more "refractory" and more "fusible" monomers leads to the melting temperature T_0.

Thus, the secondary structure of a protein has the geometry of a helix or a β-fold with an elementary unit (i.e., a turn of a spiral, or a "hairpin") that includes some three-to-ten monomers of the chain.

Meanwhile, the tertiary structure is the way the chain is laid out as a whole, that is, the geometry in which the pieces of the secondary structure are brought together. A tertiary structure is intrinsically different from the secondary one. When the secondary structure is formed, only the monomers that are close to each other along the chain are brought together. On the other hand, the formation of tertiary structure may bring close to each other any parts of the chain, even those separated by very long strands of the chain.

The two examples in Plate 4 are the tertiary structures of two proteins called recombinant histone rHMfB (a) and HIV-1 matrix protein (b), but their names are not really important for us at the moment. The spirals show the strands of α-spirals, and the flat arrows represent the pieces that have β-structure. This way of depicting the structure was suggested by the American biophysicist Jane Richardson and is good for its clarity. If you tried to draw the secondary structure in more detail, the picture would appear too complex and hard to understand.

The proteins in Plate 4 are globular ones. This means that their tertiary structures fold into a dense, compact bundle called a globule. We shall talk in detail about polymer globules in Chapter 8.

4.7 — Globular Protein Enzymes

Quite a lot of proteins have a globular structure. Above all, these include enzymes that catalyze all kinds of chemical reactions in a living cell, in particular biosynthesis of new proteins and DNA. Remember that a catalyst is a substance that speeds up a chemical (or some other) reaction, but is not itself affected by the reaction. A light-hearted example is a subway escalator. Its function is to take passengers up and down. Let us think of these two operations as of two "reactions" going in opposite directions:

$$[\text{person in subway}] + [\text{electrical energy}] \Leftrightarrow [\text{person above ground}]$$

From the point of view of energy conservation, it is all very simple and straightforward. If a person is in the subway and there is enough electrical energy available, then he *can* be moved to the surface (the direct "reaction"). Alternatively, if the person was on the surface and is going down, then his potential energy *can* be transformed into electrical energy (the reverse "reaction"). Of course, the latter process does not occur unless you have a specially constructed escalator. The escalator itself would not be affected by taking passengers up and down. This is exactly what a typical catalyst does. (You can come up with many more such examples for yourself. In fact, any kind of machine tool acts as a catalyst.) By the way, biological enzymes behave more like man-made machines, rather than like ordinary chemical catalysts (such as, say, a platinum powder that speeds up the oxidation of sulphur dioxide to sulphur trioxide by nearly 1000 times and is used in the industrial production of sulphuric acid).

There are two similarities between enzymes and machines. First of all, the acceleration of the reaction is extremely high. Usually the reaction does not even occur without the enzyme, in the same way as a rod does not spontaneously shape itself into a bolt without a lathe! The second point in common is the extreme selectivity. An enzyme may work with one substance, but would not work with another, even a very similar one. It is like a cutting tool that cuts right-handed bolts, but would not make left-handed ones.

So how do these molecular machines, the enzyme globules, actually work? Figure 4.13 shows the mechanism schematically. A "starter" molecule is to undergo the treatment. It dives into a special cavity, or a pocket, on the surface of the globule. Inside the pocket, the molecule presses a kind of "button," which is called an active center. As a result, the electron shells of the active center are set into fast motion; then other parts of the globule start moving (although not as fast).

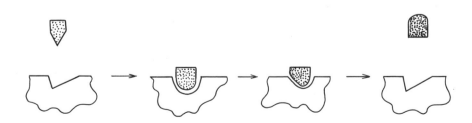

FIGURE 4.13
A sketch of different stages of catalysis by an enzyme.

They squeeze the "starter" molecule as if with a pair of pincers and pull it, snap it, wring it, and so forth, to make it into the desired shape. In a similar fashion, other proteins fight the "invaders" of our body, such as bacteria and viruses; these proteins are capable of highly specific recognition of other molecules. A realistic model of such a "recognition molecular machine" is shown in Plate 5.

Certainly, our description of enzymes and immunoglobulines is rather approximate. On the other hand, a detailed theory of how these proteins really work has not yet been completed. This study forms a subject called enzymology. In any case, what seems apparent at this stage is that since each tool cr machine is not just a random pile of bits and pieces, similarly an enzyme's globule should be an organized structure, with all the monomers in well-defined places.

This conclusion can be easily tested by experiment. If it is true, then all the molecules of a particular protein should be globules of a strictly identical shape. Therefore, if they are fairly concentrated they would line up in a regular periodic lattice (as in Figure 4.14), a protein crystal. This is exactly what happens. If you extract a protein from a cell and make a concentrated solution, then after some skillful work, you end up with protein crystals. They are so perfect that they can provide sharp diffraction patterns when illuminated with X rays. Studying such diffraction patterns is a good way to find out the spatial structure of the globules. This is just what the scientists do; as a matter of fact, hundreds and hundreds of tertiary protein structures have already been "decoded" by now.

So what are the forces that hold a protein in the shape of a globule? The globule must be very dense since all the monomers have to occupy fixed positions. It turns out that the shrinking of the protein molecule to a globule is mainly caused by the hydrophobic effect already discussed. About a half of all the 20 amino acid

FIGURE 4.14
A sketch of a protein crystal.

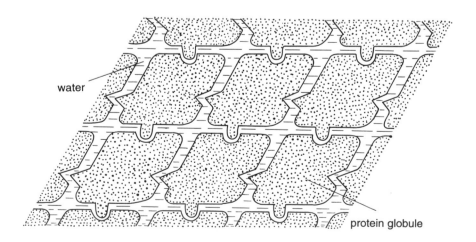

water

protein globule

residues are hydrophobic, so they are crammed into the inside the globule, letting the hydrophilic ones take up positions on the surface. This arrangement reminds us of spherical micelles (Figure 4.4a). The only difference is that now we have only one long chain strung back and forth through the globule. Thus the structure is not quite spherical, but much more complex.

We can now understand the difficulties of protein engineering. The aim is to design the primary structure so that the chain will coil up into a globule with the required tertiary structure and will act in the desired way. It is no easier than to write a decent book on polymer physics in a language that seems like gibberish, just by manipulating letters and words randomly!

4.8 — Tertiary Structures of Other Biopolymers

The tertiary structure of a globular protein, as we have seen, has quite a rigid spatial layout. There is a similar, yet less clear situation with RNA. However, the tertiary structure of DNA has been studied even less. This is because DNA molecules are very long and form spatial structures not just by themselves, but together with other proteins (whose size is far smaller than that of DNA). We shall talk a little bit more about this in Chapter 8.

4.9 — Physics and Biology

There is just one more question we would like to broach to conclude this chapter. We have been using the words: primary, secondary, tertiary. And what comes next? Sometimes the name quaternary structure is introduced when a few protein globules are stuck together, or when one protein chain forms a number of little globules. Clearly, there is the whole hierarchy of structures: There are complexes of chains, these complexes form parts of cells, the cells make up tissues, and so on.

Where does the dividing line between physics and biology go, in the face of such a variety of structures and systems? Physicists and chemists study atoms and molecules. Biologists dig their way toward them, from the other end, by looking at organs, tissues, and cells. Do their paths ever overlap, or even meet? Or are they separated by an unbridgeable gulf? This is a crucial question. If there is a gulf, then we will never succeed in understanding life. Biopolymers, with their three levels of structure (i.e., primary, secondary, and tertiary), could bridge the gulf. On one hand, a biopolymer chain is just a molecule. In this sense, it should be the job of physics to explore its properties. On the other hand, a biopolymer molecule has aspects that could be called "specifically biological."

Indeed, what do all living things, from an elephant to a microbe, have in common? One of the main features is some kind of "shape" or "construction" that the creature holds from birth to death. We could say the same about a biopolymer chain (although its design is much simpler). A biopolymer keeps its structure unchanged, from the moment it is synthesized until it is destroyed. The Russian physicist I. M. Lifshitz suggested this analogy in 1968, in describing the "linear memory" of biopolymers—as if they always "remember" the linear structure they were given when synthesized.

Thus, it turns out that if you start to look into the physics of systems with linear memory, you may hope one day to come face to face with the mysteries of biology. We shall talk more about this in the chapters that follow.

The Mathematics of a
Simple Polymer Coil

It was customary for wealthy pirates to start with a proper wardrobe.
Mark Twain, *Tom Sawyer*

5.1 — Mathematics in Physics

In the two previous chapters we looked at the properties of real polymeric substances. We have come across both artificial polymers, which are used in industry or in everyday life, and natural polymers, the building bricks of life. We only used words to describe them, without any mathematics. However, it was more like a story than a theory, so our description was rather superficial. To understand polymers better, as always happens in physics, we must move on from words to mathematics. This is because "those who have mastered at least the principles of mathematics give the impression of people with one more sense than other mortals" (Charles Darwin). Moreover, "mathematics is the language in which the gods talk to people" (Plato).

However, mathematical descriptions have their own "game plan." Real systems are so extraordinarily complex that if you wanted to describe them fully, you would have to take into account an incredible number of different factors. This would be a hopeless task. The way out is to simplify reality, grasping the main features and ignoring all the less important. Fortunately, constructing a theory

65

with even a very simple model usually pays off. When you get a deep feel for the properties of the simple model, it opens your eyes to the behavior of the real system too.

In this chapter, then, we shall discuss different mathematical descriptions of the simplest model of a polymer, the "ideal polymer coil" (the reason for this name will become clear in Chapter 7).

To tell the truth, the authors are physicists, not mathematicians. So we can fully appreciate Goethe's joke when he said: "Mathematicians are a bit like the French—whatever you say to them, they will immediately translate it into their own language and it will become something totally different!" Of course, Goethe's sarcasm was directed against the invading armies of Napoleon, rather than against the poor mathematicians. However, his attitude is echoed by John Ziman, the present-day English theoretical physicist: "Nothing is more repellent to normal human beings than the clinical succession of definitions, axioms, and theorems generated by the labours of pure mathematicians." Therefore, we shall try to spice up the following chapters with some history and various physical analogies. Occasionally we may even wander off into some "non-polymer" physics. In any case, we are not going to do math just for the sake of it. The physical sense and meaning of mathematical formulas will be our main concern.

5.2 — Analogy between a Polymer Chain and Brownian Motion

Imagine you are in a thick forest. You have picked enough mushrooms and berries (or whatever you were gathering there), the weather has become bad, and all you want now is to get out of the wretched place. But how? The trees and bushes hinder your view and make it hard to walk. You cannot see the sun behind the clouds. If you do not have a compass, it seems quite clear that you will have a hard time, at least in theory. (Well, in practice they say that an experienced person can tell directions by looking at how moss and lichen grow on tree trunks, where ant hills are, etc.) On the other hand, even if you had a compass, would it be of any use without a map? You would not know which direction you need to take. Even so, it appears a compass would still be extremely useful. Soon we shall see why.

We are telling you this story for a good reason—to help you comprehend a deep mathematical concept that has been very fruitful when explaining the behavior of polymers. Historically, this concept was first developed when Brownian motion was studied.

Brownian motion was discovered in 1827 by the English botanist Robert Brown (1773–1858). Looking through a microscope at little particles of pollen suspended in water, he was fascinated by their random "dances." The particles were moving by themselves, with no external encouragement. So a lot of people decided that there must have been some "living power" causing the motion (because the flowers were animate!). They reckoned this had proved that there was some mysterious "substance" that made the animate different from the inanimate.

The question was debated for a long time. Everyone was free to think what they wanted. Then there was a dramatic boom in interest at the end of the 19th century. Brownian motion was regarded as a kind of perpetual motion, so it tantalized those who were puzzled by the general problems of science. Such problems included the nature of irreversibility (i.e., the distinction between past and future) as well as the difference between Darwinian biological evolution leading to the perfection of species and the thermodynamic evolution described by Clausius, Thompson, and Boltzmann, which leads to dissipation, or, as it was then called, "thermal death."

Eventually, the answer was found by Albert Einstein[1] and the Polish physicist Marian Smoluchowski (1872–1917), a professor at the University of Lviv. The title of one of Einstein's papers on the theory of Brownian motion is rather telling: "On the motion of particles suspended in resting water which is required by the molecular-kinetic theory of heat." Einstein and Smoluchowski considered chaotic thermal motion of molecules and showed that this explains it all: A Brownian particle is "fidgeting" because it is pushed by a crowd of molecules in random directions. In other words, you can say that Brownian particles are themselves engaged in chaotic thermal motion. Nowadays, science does not make much distinction between the phrases "Brownian motion" and "thermal motion"— the only difference lies back in history. The Einstein–Smoluchowski theory was confirmed by beautiful and subtle experiments by J. Perrin (1870–1942).[2] This was a long-awaited, clear, and straightforward proof that all substances are made of atoms and molecules.[3]

We will skip further details of this adventure story. We just need to stress one more thing before we get back to polymers. Since a Brownian particle moves due to collisions with molecules, its path breaks into a sequence of many short flights

---

[1] By the way, Einstein presented his theory of relativity and the concept that light consists of photons in exactly the same year, 1905.

[2] You can read about Perrin's experiments in a very interesting book by G. Trigg: *Crucial Experiments in Modern Physics* [7].

[3] The atomic hypothesis was suggested long ago by the ancient Greeks, but it had to wait for more than two thousand years to be proved!

and turns. In this sense, a Brownian trajectory is pretty similar to the shape of the polymer chains we saw in Section 2.4 (Figure 2.6). Another obvious example of this sort is of a man who is lost in a forest, with no map, or compass, and has no choice but to wander at random.

Certainly, no microscope would let you see the twists and turns of an individual molecule's path. However, the Einstein–Smoluchowski theory tells us how to spot the difference between a "fuzzy" line, which consists of a great number of tiny random kinks, and an ordinary smooth curve, even though we cannot discern the individual kinks. (We do not always need to see everything; for example, we can happily tell water from alcohol even though the individual molecules are invisible!) In the same way, a polymer chain looks nothing like a shape stretched in a certain direction. And the path of a man in a forest would depend quite noticeably on whether he is equipped with a compass or not!

So what is the difference between a smooth and a "kinky" path?

For motion in a straight line: $R = v(t_2 - t_1)$ (5.1)

For a Brownian particle: $R = \ell^{1/2}[v(t_2 - t_1)]^{1/2}$ (5.2)

The notation here is as follows: R is the displacement, the distance between the points where the moving particle was at the times t_1 and t_2 ($t_1 < t_2$); v is the average velocity of the motion. In formula (5.2), ℓ has the dimension of length (we shall explain its physical meaning a little later).[4] Meanwhile, R is the mean (or rather root-mean-square) displacement, i.e., $R = \langle(\vec{R}_2 - \vec{R}_1)^2\rangle^{1/2}$ where $\vec{R}_1$ and $\vec{R}_2$ are radius vectors showing the particle's position at the moments t_1 and t_2. The angle brackets indicate that the average is taken over a number of different Brownian paths.

What is the polymer analogue of the Einstein–Smoluchowski equation (5.2)? Let L be the contour length of a polymer chain. It is bound to be proportional to the number of monomers in the chain, given that the chemical structure does not change. The chain length L plays the same role for a polymer as the value $v(t_2 - t_1)$ for a Brownian particle, that is, the total distance traveled by the particle along the path. Since the chain wiggles around a lot, the root-mean-square distance between its ends, $R = \langle(\vec{R}_2 - \vec{R}_1)^2\rangle^{1/2}$, is totally different and not even proportional to the contour length L. You can easily find R from equation (5.2) if you replace $v(t_2 - t_1)$ by L:

$$R = \ell^{1/2}L^{1/2} = (\ell L)^{1/2}.$$ (5.3)

---

[4] The Einstein–Smoluchowski theory leads to the value $\ell = mkT^{1/2}(3\pi\eta r)$ for spherical Brownian particles of radius r and mass m moving in a liquid of viscosity η at a temperature T.

Brownian motion is shown in the special movie called *"Brownian Motion"* on the CD ROM. If you measure carefully how far the particle walks on the screen of your computer, you should be able to recover the "square root law" (equation (5.2)).

5.3 — The Size of a Polymer Coil

Let us look again at formulas (5.1) and (5.2). How different are they really? And how similar are the end-to-end distance R and the contour length L of a polymer coil? The answer is that the differences are essentially large, and we shall try to explain why.

The main distinction between (5.1) and (5.2) can be spotted at the first glance—the power law for R depending on the time interval $\Delta t = t_2 - t_1$ is not the same! In case you think that is not important, or not even worth mentioning, we will use words and numbers rather than formulas to make it clearer.

Let's start with numbers, in other words, with estimating the order of magnitude of things. As we have seen, the rambler lost in a forest is very similar to what happens to a Brownian particle. Say the rambler spends 10 hours per day walking (i.e., $\Delta t = t_2 - t_1 = 10$ hr) at a speed of $v = 3$ km/hr. (It is hard to move much faster in the forest.) If he uses a compass or some other means of judging direction, his path will look like a straight line (at least on average). His displacement will be given by (5.1); in our case, it is 30 km. Having strolled that far, the well-prepared hiker is quite likely to find the way out, or at least to reach a major road or path. However, if he wanders randomly with no guidance, the situation becomes much more serious. Let's just take for granted for a moment that $\ell \approx 300$ m; then we can use (5.2) and discover that the poor chap will move no more than 3 km from where he started. Hence, good advice for stray hikers in a forest is: Do not rush about! Just keep going in a fixed direction. It does not matter which direction, as long as you use some landmarks to guide you in a straight line.[5]

---

[5] Even if the forest is huge, and you know that one of its borders is much closer than the opposite one, you still should not wander around, but advance in a straight line. Only if the lapse of time suggests that you must be straying away from the closest border should you turn abruptly (e.g., through 120°) and try another direction. This sounds like a lovely problem for math enthusiasts—to work out the best strategy for a rambler who starts to make his way out of a forest at a given distance from the closest straight-line border.

By the way, lost travelers are often known to feel that they are going in circles. All kinds of absurd causes have been suggested to account for this. Some reckon it is because one's two legs may differ in length or strength, others blame Coriolis forces, or even giddiness due to the rotation of the Earth. However, the proper scientific explanation immediately follows from Figure 2.6. A meandering, entangled path tends to cross itself a number of times, so it is not surprising that a lost traveler may occasionally find himself in a place where he has been before. In Chapter 7, we shall estimate how many self-crossings, on average, there are along a random path.

Now, what does a "self-crossing" mean in the case of a polymer? The entangled shape brings some monomers, which are separated by a long piece of the chain, closer together. They may even collide. Interactions between such monomers during their collisions are known as volume interactions. They occur in the bulk of a polymer coil, in contrast to "linear" interactions that hold together the neighboring monomers along the chain. A little later, we shall discuss in detail how various properties of polymers are affected by volume interactions.

Let's go back to formulas (5.1) and (5.2), and look at another example directly connected with polymers. You may remember that the contour length of a DNA double helix from a human or an animal cell can be as big as one meter. The question is, how small a coil would the double helix form if it meandered randomly in a Brownian-like way? The value of ℓ for DNA has been measured in experiments; to high accuracy, $\ell = 100\,\text{nm} = 10^{-7}\,\text{m}$. Thus, equation (5.3) leads to

$$R \approx 3 \cdot 10^{-4}\,\text{m} = 0.03\,\text{cm}. \tag{5.4}$$

This is certainly much less than the one-meter contour length, yet far too big to squeeze into a cell nucleus, which is about $10^{-6}\,\text{m}$ in diameter! Another reason why DNA's shape cannot be merely random is that there has to be some regularity in it; otherwise it would be impossible to find its different strands and "read" the information from them. Scientists are still a bit vague on exactly how the double helix is packed into a cell (see Plate 2). There are some ideas, however, which we are going to discuss in Chapter 8. For now we shall only say that the main part is played by the volume interactions mentioned above. They help DNA to fold in such a way that its size becomes proportional to the number of monomers N (i.e., the contour length L) not raised to the power 1 (as for a stick), or even to the power 1/2 (as for a randomly entangled coil described by (5.2) or (5.3)), but to the even smaller power 1/3.

5.4 — Derivation of the Square Root Law

Now we know that the main peculiarity of a randomly entangled polymer coil is that its size is proportional to the square root of the chain's contour length: $R \sim L^{1/2}$ (see (5.3), for example). This is why a polymer molecule appears much smaller in size than it would be if entirely stretched (given that it is reasonably long). Indeed, the ratio $R/L \sim 1/L^{1/2}$ decreases with L, and tends to zero when $L \to \infty$ for any ℓ. However, if we wanted to find the size R of a particular molecule with a given contour length L, we would need to know the value of ℓ. In any case, it would help if we knew more about the physical meaning of ℓ. Also, since we realize the importance of the square root law, it might be a good idea to see where it comes from.

To tackle both questions, we shall abandon the Brownian analogy and return to the purely polymer world, as it will be easier that way. Let's imagine a freely jointed polymer chain consisting of N units (Figure 2.4). The end-to-end vector $\vec{R}_N$ is given by a simple formula, which is obvious from Figure 5.1:

$$\vec{R}_N = \sum_{i=1}^{N} \vec{r}_i. \tag{5.5}$$

Here i is a number of a current chain segment, N is the total number of segments in the chain, and $\vec{r}_i$ is an end-to-end vector for the ith segment. The moduli of all the vectors $\vec{r}_i$ are the same; they are equal to the length of one segment: $|\vec{r}_i| = \ell$ (we

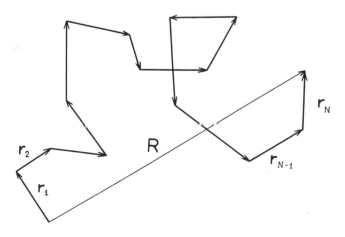

FIGURE 5.1
Illustration for formula (5.5).

have deliberately chosen this notation for the length of a segment). Meanwhile, the directions of the vectors $\vec{r}_i$ are completely random and independent of each other.

Thus, the value of $\vec{R}_N$ that describes the size of the coil can be written as the sum of a large number of independent random terms, as in (5.5). Fortunately, the mathematical properties of such sums have been very well studied since they often emerge in areas of math, physics, engineering, and biology.

We will pick just a couple of examples. The stress experienced by an airplane, for instance, depends on the total weight of the passengers, the fat and the skinny altogether, and no one knows whether possible deviations from the average weight are important. Similarly, in light scattering, the electromagnetic field of the scattered wave is the superposition (i.e., the sum) of the fields created by individual atoms. These contributions are totally random due to thermal motion of the atoms; when they interfere they can either enhance or diminish each other. In both examples, as well as in many others, the answer about possible deviations from the average can be found from the square root law. Let's derive this law from formula (5.5).

Together with $\vec{R}_N$, we need to introduce $\vec{R}_{N-1}$, which is a similar end-to-end vector but for the the first $N-1$ units of the chain. We can write

$$\vec{R}_{N-1} = \sum_{i=1}^{N-1} \vec{r}_i, \qquad \vec{R}_N = \sum_{i=1}^{N} \vec{r}_i$$

$$\vec{R}_N = \vec{R}_{N-1} + \vec{r}_N \tag{5.6}$$

(Recursion relations of this kind are often useful when trying to sort out random values.)

Now we can start thinking about how to work out the average end-to-end distance. But first we need to decide what sort of average to look at. The point is that the average value of the vector $\vec{R}_N$ itself, as well as of all its components, is zero; that is, $\langle \vec{R}_N \rangle = 0$. This is simply because the end-to-end vector can be equal to $\vec{R}_N$ and to $-\vec{R}_N$ with the same probability. Therefore, it is the average length of the vector, $\langle |\vec{R}_N| \rangle$, that is meaningful and that gives us an idea of the size of the coil. However, it is handier to calculate the value

$$R_N^2 \equiv \langle \vec{R}_N^2 \rangle = \langle \vec{R}_N \cdot \vec{R}_N \rangle = \langle |\vec{R}_N|^2 \rangle, \tag{5.7}$$

which also describes the coil's size. The definition (5.7) agrees with the way we defined R before.

According to (5.6), R_N^2 is given by:

$$\vec{R}_N^2 = \vec{R}_{N-1}^2 + 2\vec{R}_{N-1}\vec{r}_N + \vec{r}_N^2 = \vec{R}_{N-1}^2 + 2|\vec{R}_{N-1}|\ell\cos\gamma_N + \ell^2, \quad (5.8)$$

where γ_N is the angle between vectors $\vec{R}_{N-1}$ and $\vec{r}_N$, and ℓ is the modulus of $\vec{r}_N$ (i.e., $\ell = |\vec{r}_N|$). In the case of a freely jointed polymer, the direction of $\vec{R}_N$ does not depend on the shape of the rest of the chain. This is why the angle γ_N is equally likely to have any value from $0°$ to $180°$, which means that $\cos\gamma_N$ is just as likely to be positive (when γ_N lies between $0°$ and $90°$) as negative (for γ_N between $90°$ to $180°$). Therefore, the average value of the cosine is zero, $\langle\cos\gamma_N\rangle = 0$. This should help us to find the average value of $\vec{R}_N^2$ straightaway from (5.8). Indeed, since the average of the second term on the right-hand side is zero,

$$\langle\vec{R}_N^2\rangle = \langle\vec{R}_{N-1}^2\rangle + \ell^2. \quad (5.9)$$

So, if an extra segment is added to the chain, $\langle\vec{R}^2\rangle$ will increase by ℓ^2. Now, applying induction, we can easily prove that

$$\langle\vec{R}_N^2\rangle = N\ell^2 = L\ell. \quad (5.10)$$

Now, at last, we can use equation (5.7) to find the size of a polymer molecule consisting of N units:

$$R_N = \langle\vec{R}_N^2\rangle^{1/2} = N^{1/2}\ell = L^{1/2}\ell^{1/2}. \quad (5.11)$$

Thus, the "$L^{1/2}$ rule" is proved.

5.5 — Persistent Length and Kuhn Segment

We have proved equation (5.11) only for a particular model of polymer, with independent, freely jointed segments. Is the formula valid for other models (including random walks)? We would need a special investigation to find out. The investigation, however, can be reduced to a very simple argument.

As we know, the flexibility of a polymer chain is not very noticeable at smaller scales, but it starts showing up as the scale increases. This means that there has to be some critical length for each polymer, ℓ_{eff}. Any segment shorter than ℓ_{eff} can be regarded as rigid; that is, its end-to-end distance is roughly the same as its contour length. At the same time, different segments of length ℓ_{eff} behave as independent. Such segments of length ℓ_{eff} are called effective segments, or Kuhn segments, after the German-Swiss physical chemist Werner Kuhn (1899–1963), who was the first to suggest this idea. Obviously, a molecule of contour length L contains $N_{eff} = L/\ell_{eff}$ Kuhn segments. Since Kuhn segments are nearly independent, we can imagine that they are freely jointed and can use equation (5.11):

$$R^2 = \langle \vec{R}^2 \rangle = N_{eff}\, \ell_{eff}^2 = \left(\frac{L}{\ell_{eff}} \right) \ell_{eff}^2 = L\ell_{eff}. \qquad (5.12)$$

Equation (5.12) gives, in fact, the definition of effective length; that is, $\ell_{eff} = R^2/L$. Comparing (5.12) with (5.3), we immediately discover that the value ℓ that appears in (5.3) and (5.2) is exactly the Kuhn length. It gives the length scale on which the polymer chain (or the path of a Brownian particle) remains roughly a straight line. (This is precisely where our estimate for the lost hiker in a forest, $\ell \leq 300$ m, comes from.)

There is quite a large range of Kuhn lengths for real polymers. Rather modest values of about 1 nm are typical for simple synthetic chains, whereas DNA's effective segment stretches 100 nm. (This is a huge number, considering that an atom's size is about of 0.1 nm!)

Why is there such a difference? In each case, ℓ_{eff} is determined by the flexibility of the chain. It might be interesting to see exactly how the two quantities are linked together. We journey along the chain (Figure 5.2), assuming that the direction of the first segment is fixed. At first, the change in direction would be very smooth and hardly noticeable. It feels as if the chain keeps a sort of "memory" of the initial direction. Farther on, this memory starts fading, and then completely disappears. To describe this quantitatively, we choose two points on the chain, separated by a segment of contour length s (Figure 5.2). Since the chain flexes, its directions at the two points are different; let's say the angle between them is $\theta(s)$. This angle varies due to fluctuations (i.e., due to thermal motion). You could probably guess that a meaningful value is the average $\cos \theta(s)$. It turns out that, if s is reasonably

FIGURE 5.2
Diagram explaining the concept of persistent length.

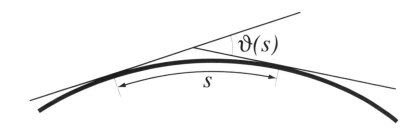

large, $\langle \cos \theta(s) \rangle$ decays exponentially with s,

$$\langle \cos \theta(s) \rangle = \exp \left(-\frac{s}{\ell} \right) \tag{5.13}$$

This formula is actually an exact definition of a very important quantity, ℓ, which is often known as the persistent length of a polymer chain.

What is the physical meaning of the relationship (5.13)? To answer this, let's first look at a segment that is shorter than ℓ. When $s \ll \ell$, equation (5.13) leads to $\langle \cos \theta(s) \rangle \approx 1$. Hence the angle $\theta(s)$ fluctuates around zero. This simply means that chain segments that are close compared to ℓ have nearly the same direction. For the opposite case, $\ell \gg s$, equation (5.13) results in $\langle \cos \theta(s) \rangle \approx 0$. Clearly, this indicates that $\theta(s)$ can be anything from $0°$ to $360°$ with equal probability. So the chain direction gets totally "forgotten" at lengths greater than ℓ.

To summarize, persistent length is a parameter describing polymer chains quantitatively. Its physical meaning is the following: Memory of chain direction is retained on length scales shorter than ℓ, but lost once ℓ is exceeded.

Since the memory stretches in both directions, the Kuhn segment of length ℓ_{eff} must be roughly twice as big as the persistent length ℓ. This is indeed true. Moreover, the relationship $\ell_{eff} = 2\ell$ is exact for a wormlike polymer chain (see Section 2.3); it is also valid for other models, although only approximately.

In principle, the persistent length should vary with temperature. The higher the temperature of a chain, the more it bends, and hence, the shorter its persistent length and Kuhn segment. However, in most cases this dependence is not important, since the range of temperatures where polymers may even exist is not that wide.

5.6 — The Density of a Polymer Coil and Concentration Ranges of a Polymer Solution

The phrase *polymer coil* may perhaps remind you of a ball of string. In some sense, string is indeed rather similar to a polymer chain. Despite this, a ball of string and a polymer coil have nothing in common whatsoever. A ball is wound tightly, with no gaps, whereas a polymer chain is arranged in a very loose manner, as Figure 2.6 showed. Yet structures as tight as a ball *are* known in the polymer world—they are called globules. We shall leave them until Chapter 8, and instead look for a theory to explain such low densities of polymer coils.

As we have seen, the size of a molecule made of N effective segments, of length ℓ each, is equal to $R = \ell N^{1/2}$ (see (5.12)). The volume of this coil can be estimated as[6] $V \sim (4/3)\pi R^3 \sim R^3 \sim \ell^3 N^{3/2}$.

Knowing the volume V and the number of segments N, we can now find the average concentration (number of segments per unit volume), $c^\star$, of the segments in the coil:

$$c^\star = \frac{N}{V} \sim \frac{N}{\left(\ell^3 N^{3/2}\right)} = \ell^{-3} N^{-1/2}. \tag{5.14}$$

Our main achievement so far is that we have found the actual dependence of $c^\star$ on N. In particular, estimate (5.14) shows that, if a chain is sufficiently long ($N \gg 1$), the segment concentration $c^\star$ becomes extremely low.

You might remember that we came across the concentration $c^\star$ in Section 3.6, when we talked about polymer solutions. There we said that if the overall concentration c of a polymer in the solution is less than $c^\star$ ($c < c^\star$), then individual coils hardly ever overlap and the solution looks like a low pressure gas of coils (Figure 3.7a). If, on the other hand, $c > c^\star$, then the coils penetrate deeply into each other and the chains are entangled (Figure 3.7c). The value $c^\star$ corresponds to the threshold regime (Figure 3.7b). Now we have seen that the value $c^\star$ is extremely small for long polymer chains. It means that a polymer solution can be made up of separate, nonoverlapping coils only at very low concentrations.

This conclusion becomes even clearer if we replace c by another quantity, the volume fraction ϕ of a polymer in the solution. Let v be the volume of a single segment, with c such segments per unit volume. Hence, the fraction of the whole volume occupied by the segments is $\phi = cv$. The advantage of ϕ (compared to c) is that it has no dimensions. Formulas become even easier—for example, in the case of a polymer melt (when there is no solution at all), $\phi = 1$.

Let d be a characteristic thickness of a polymer chain. Then, thinking of an effective segment as a cylinder of diameter d and height ℓ, we can estimate its volume: $v \sim \ell d^2$ (we leave out as usual an unimportant factor of $\pi/4$). The threshold volume fraction $\phi^\star$ that marks the transition between Figure 3.7a and Figure 3.7c will be approximately the following:

$$\phi^\star \sim c^\star v \sim \left(\frac{d}{\ell}\right)^2 N^{-1/2}. \tag{5.15}$$

---

[6] The sign $\sim$ means "of the same order of magnitude." When we make rough order of magnitude estimates, such factors as $4/3\pi$ are not important, so we leave them out.

In practice, the ratio ℓ/d ranges from numbers of order 1 (2–3) for flexible synthetic polymers to 50 for a DNA double helix. Using (5.15), we can look at particular numbers. For example, if there are 10^4 units in a flexible chain, then the coils already start overlapping when the volume fraction of the polymer is 10^{-2}, in other words, when the polymer takes as little as 1% of the whole volume of the solution.

Therefore, if $N \gg 1$, there has to be a rather broad range of concentrations $\phi^* < \phi < 1$ ($c^* < c < a^{-3}$) in which the coils are heavily entangled ($\phi^* < \phi$), yet there is still little polymer in the solution ($\phi < 1$). This type of polymer solution is known as semidilute and is shown in Figure 3.7c. On the other hand, if $\phi \sim 1$, that is, the volume fractions of the polymer and the solvent are comparable, the solution is called concentrated (Figure 3.7d).

By the way, the intermediate, semidilute region is only possible because polymer chains are so extremely long ($N \gg 1$). Indeed, if $N \sim 1$ (i.e., if there were small molecules in place of a polymer in the solvent), the two inequalities $\phi^* \simeq N^{-1/2} < \phi < 1$ would not work together. Therefore, the solution has to be either dilute ($\phi < 1$), in which case the individual molecules hardly interact with each other, or concentrated ($\phi \sim 1$), with strongly interacting molecules.

Figures 3.7a–3.7d depict isotropic polymer solutions in which there is no preferential orientation of polymer chains. However, as we have already mentioned, spontaneous ordering can occur in concentrated solutions of rigid ($\ell/d \gg 1$) polymers, and they become liquid crystalline (Figure 3.7d). In most cases, such an anisotropic state is established at concentrations $c > c_{cr}$, where $c^* < c_{cr} \sim \ell^{-2}d^{-1}$, so c_{cr} corresponds to a semidilute solution.

**There is a special movie on the CD ROM called *"Polymer Solution."*
In the first part, it shows a dilute solution, where you can see coils
moving far away from one another. In the second part, several
additional coils are added, the concentration gets higher, and coils
start to penetrate and entangle with each other, thus making a
semidilute solution. Different coils are shown in different colors to
make it easy to keep track of each of them.**

5.7 — The Gaussian Distribution

There is another type of problem that we still need to tackle when portraying an isolated polymer coil. We might as well start with a very simple question: What does it mean to say that a coil size is proportional to the square root of the chain length, i.e., $L^{1/2}$ or $N^{1/2}$? Can the chain, accidentally, stretch out into a straight line? In the same spirit you may ask, can one day all the passengers of an airplane turn up very fat, and the plane fail to take off? Can the dipoles of an object that scatters light, by coincidence, all line up in the same direction?

You would probably agree that, in principle, all these things could happen, although they are extremely unlikely. In fact, the probability that a chain would stretch out into a straight line is just the same as the probability of any other particular configuration. But the whole point is that there are many, many curled up and entangled shapes, whereas there is only one straight line and no more. That is why a polymer, left on its own, is most likely to roll up into a coil of size $R \sim \ell N^{1/2}$. Hence, this is just what the average size of a polymer is, within an order of magnitude. The chances that due to fluctuations a polymer may expand up to $R \sim \ell N$ are exceedingly slim.

To bring some math into play, we need to count up all the stretched and all the coiled conformations of a polymer chain. More precisely, we would like to know how many different conformations of the chain have the same end-to-end vector, $\vec{R}$. How much is this as a fraction of all the possible conformations? (In other words, what is the probability that a polymer chain picked at random has an end-to-end vector $\vec{R}$?) It is not a usual kind of task for elementary math. The question really is, in how many different ways can you choose the terms of a sequence so that their sum stays the same?

The derivation is too complex to give here, so we shall merely tell you the answer:

$$P_N(\vec{R}) = Q \exp\left(-\frac{3\vec{R}^2}{2N\ell^2}\right). \tag{5.16}$$

Here ℓ is the effective, or Kuhn, length; N is the number of Kuhn segments in the chain; and Q is a constant factor. The value of Q depends on what exactly we mean by $P_N(\vec{R})$. It can be two things: either the total number of conformations with the given $\vec{R}$ (in some infinitesimal volume element, of course) or the probability of a conformation with this $\vec{R}$ (i.e., the ratio of the number of such conformations to the total number of all the possible conformations). In particular, if P_N is the probability, then

$$Q = \left[\frac{3}{2\pi N\ell^2}\right]^{3/2}. \tag{5.17}$$

Equation (5.16) is written for some unknown vector $\vec{R}$. We may also want to look at its components R_x, R_y, and R_z (which show how far apart the two ends of the chain are shifted from one another along the x-, y-, and z-axes). For any of the three components:

$$P_N(R_\alpha) = \left[\frac{3}{(2\pi N \ell^2)}\right]^{1/2} \exp\left(-\frac{3R_\alpha^2}{2N\ell^2}\right). \qquad (5.18)$$

(Formula (5.16) can be obtained by multiplying together expressions of type (5.18) for all the three components $\alpha = x$, $\alpha = y$, $\alpha = z$; we leave it for you to check and explain.) The function $P_N(R)$ is plotted in Figure 5.3. You can see that all the values of R from zero up to about $\ell N^{1/2}$ have roughly the same probability. However, if R becomes larger, the probability decays very dramatically. We could phrase it like this: If the first unit of the chain is fixed at the origin, then there is a roughly equal chance that the last unit will be at any point inside a sphere of radius $\ell N^{1/2}$. On the other hand, the likelihood of finding the last unit outside the sphere is negligibly small.

Equations (5.16) and (5.18), for the probability of different values of $\vec{R}$ or its components, are known as Gaussian distributions, because famous mathematician K. F. Gauss (1777–1855) was the first to come up with this kind of formula, albeit in a different context. The number of situations in which one encounters Gaussian distribution is amazing. For example, as you probably know, to obtain more accurate data and to reduce the impact of inevitable errors in measurements, experimentalists must repeat the same measurement again and again. Let's say they do it n times. Then they need to find the average. To do this, they first have to add all the measurements together. Thus, the errors get added up too. However, some of the errors are positive and some are negative, at random. (It seems we can never get away from sums of random values!) This is why the error in the sum is proportional to $n^{1/2}$, rather than to n. Finally, to take the average, the sum

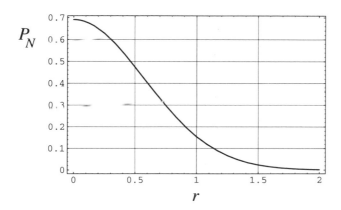

FIGURE 5.3
The dependence of P_N on $r = R_\alpha/(\ell N^{1/2})$ given by equation (5.18).

is divided by n. That is why the error in the average will behave as $n^{-1/2}$, hence it will indeed decrease with the number of measurements. In general, we can say that the sum of a great number of random values is controlled by a Gaussian distribution of probabilities, like (5.16). This is one of the key ideas in probability theory. Due to its great importance, it was given a posh name, the central limit theorem (CLT).

Why "limit"? In fact, the Gaussian distribution (5.16) is valid exactly only in the limit of a very large number of terms. For a finite number of terms, it is only approximate. However, even in the case of a moderate number of terms (or unit segments, N), a Gaussian distribution provides acceptable accuracy.[7]

Quite naturally, a free single polymer coil is often called a Gaussian coil, after the distribution (5.16).

---

[7] If you have a liking for elegant mathematical trifles, you may be interested in the following example. It is about so-called lucky bus tickets. Russian bus tickets used to have six-digit numbers on them, and Russians liked to believe that a ticket was lucky if the sum of the first three digits was the same as the sum of the last three digits. It is possible to prove that there are 55,252 lucky tickets out of the total quantity of 1,000,000 six-digit numbers. So the probability of a lucky ticket is 5.5252%. On the other hand, the sum of three digits of a number is actually a sum of random terms. Although there are only three terms in this case, you could still try to use the CLT to estimate the probability of a lucky ticket. You would then get an approximate answer of 5%, which is surprisingly close to the accurate value. Thus, even if N is as small as 3, the CLT works reasonably well.

The Physics of High Elasticity

6.1 — Columbus Discovered... Natural Rubber

In some popular books, they say that natural rubber was the first polymer encountered by our ancestors. If you think about it for a minute, your reaction might be: "What rubbish! People had always known things like wood and timber. Not to mention that both prehistoric and contemporary human beings themselves are made of polymers." However, the polymeric nature of many materials, although very important, only shows up in rather subtle ways. For instance, there are low molecular weight compounds that look similar to polymers in the semicrystalline, viscous, or glassy state. However, there is a property that is both very noticeable and purely polymeric, that is, impossible for small molecules: high elasticity. And the first viscoelastic substance that the Europeans came across was indeed natural rubber.

When they reached America, the first European explorers and immigrants were, of course, overwhelmed by novelties. They found potatoes, tobacco, sweet corn, and tomatoes. They met strange local people who had an unusual way of life. It is not surprising that books on the discovery of America are so enthralling. Unfortunately, some of the newcomers were greedy and aggressive and these features were much stronger in them than the natural curiosity of discoverers. Thus many of the Europeans not only made the Native Americans suffer, but also caused irreversible damage to their culture.

The culture of making rubber was a lucky exception. People from the very first expedition of Columbus were amazed by the balls that the Native Americans played with. At that time in Europe, balls were made from bull's bladders encased in leather shells (these are also polymers, by the way!). In contrast, the American balls were solid, heavy, and surprisingly bouncy. They were made from a substance which the Native Americans called "caoutchouc." Not until ten years later, when the Europeans reached as far as Brazil, they found out that "caoutchouc" is a milky resin excreted by the tree called "heve", extracted by sloping cuts on the trunk. Then the invaders, in their bellicose excitement, forgot about the rubber tree for a while. More than two hundred years passed before "heve" was properly described (in 1738). Now it is fairly well known as *Hevea brasiliensis*, and "caoutchouc" is still the word for natural rubber in some European languages.

6.2 – High Elasticity

Rubber has very unusual properties. Under certain conditions, it remains solid (nonfluid), yet is extremely elastic. A fairly low stress is able to deform a piece of rubber quite significantly (much more than if it were an ordinary solid). The deformation is reversible (elastic); that is, when the stress is released, the sample regains its original, undeformed shape.

To appreciate the elasticity of rubber, let's see how it differs from ordinary, nonpolymeric materials. Take steel, for example. Figure 6.1 compares the dependence of the stress σ on the strain $\Delta\ell/\ell$ (equation 3.1) for a steel rod (Figure 6.1a) and for a piece of rubber (Figure 6.1b). The two graphs have some things in common. They both start off as a linear relationship between σ and $\Delta\ell/\ell$; that is, they follow Hooke's law (3.1) for small deformations. In this range, deformations are almost completely reversible. The linearity extends to point A in both pictures. Then, between A and B, both lines bend. Here the deformations become significantly nonlinear, but still remain reversible. Interesting behavior starts above B; this is where the reversibility is finally lost. The sample, as they say, starts to flow. Even if the stress is released, the material retains a certain residual

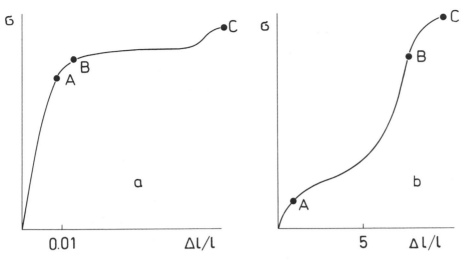

FIGURE 6.1
The dependence of
stress σ on the relative
strain $\Delta\ell/\ell$ in the case
of (a) steel and (b)
rubber.

(plastic) deformation and never goes back to its original shape. Eventually, at point C, the sample gets torn apart.

Although both graphs have points A, B, and C, they are positioned differently. There is also a huge difference in numbers. For instance, to break steel (i.e., to reach point C) you only need to stretch it by a few percent. And reversible deformations are never higher than 1%. In contrast, a rubber strip can happily extend up to eight times its original length (by 700%). Moreover, this is still reversible! The actual values of stress at breaking point are also beyond comparison. They are $2 \cdot 10^9$ Pa for steel and $3 \cdot 10^7$ Pa for rubber. So we are talking about totally different scales. This is why there is a disparity in the Young's moduli E as well. Indeed, according to (3.1), Young's modulus is determined by the slope of the linear part of a curve such as those in Figure 6.1. Hence, we get $E \approx 2 \cdot 10^{11}$ Pa for steel, and $E \le 10^6$ Pa for rubber. This gap is enormous, more than five orders of magnitude. (We have already mentioned this in Chapter 3.) One more difference is that rubber has a wide range where deformations are nonlinear yet reversible (between A and B), whereas for steel this area is almost missing. On the other hand, the curve $\sigma(\Delta\ell/\ell)$ for steel has a comparatively broad region of plastic deformations (between B and C), whereas rubber almost immediately snaps as soon as it starts to flow (Figure 6.1b).

Now you know what people mean when they talk about the high elasticity of rubber. In brief, high elasticity means that the material is prone to very high, nonlinear, yet reversible deformations as a result of rather moderate stress.

6.3 — The Discovery of Vulcanization

Besides balls, the Native Americans used to make other handy things from rubber, such as waterproof shawls, a kind of Wellington boots, flasks, and many others. They had not quite reached perfection, though. All the stuff was rather sticky and short-lasting. Even worse, on a hot day it would melt altogether! The Europeans took over and kept trying to find a better use for the unusual substance. But it was not that easy.

The problem is that natural rubber is not really a solid. Any external force (such as gravity) makes it flow slowly. Strictly speaking, it is a liquid polymer melt in the viscous state. Therefore, any natural rubber product keeps changing shape. Exactly how much it "wobbles" or "oozes" depends strongly on the temperature. High above room temperature, natural rubber is more of a liquid. At low temperatures, the oozing nearly stops. But the high elasticity disappears as well and the rubber hardens.

An amusing true-life story about the rise and fall of a man named C. Macintosh, from Glasgow, Scotland, illustrates the problem. The clever chap decided to use rubber in the production of raincoats. A thin layer of rubber was placed between two layers of fabric. It worked out very nicely, and the raincoats (called macintoshes) were very popular in the notoriously wet Britain. Macintosh rapidly became rich, shortly after he started his business in the winter of 1820. However, when summer came along, the temperature rose and all the rubber flowed out of the macintoshes. The poor inventor went bankrupt, and the whole idea of padding coats with rubber was abandoned for many years.

Not for too long though. A breakthrough occurred in 1839 when the American C. Goodyear suggested the process for vulcanizing rubber. At the molecular level, rubber consists of polymer chains with frequent double bonds (Figure 6.2). The vulcanization involves adding sulphur atoms to the rubber melt. They form covalent bonds between the chains (Figure 6.2b), so the chains become linked together by sulphur bridges. The result is a polymer network. It is not fluid, even

FIGURE 6.2
(*a*) A polymer melt before vulcanization, (*b*) a polymer network as a result of vulcanization, (*c*) a stretched polymer network.

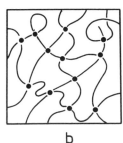
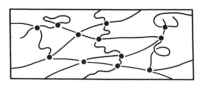

a   b   c

at relatively high temperatures when a normal polymer melt of unattached chains would start flowing (due to intensive thermal motion, making the chains move with respect to each other). At the same time, there is nothing to stop such a network from expanding. Under strain all the chains would stretch as in Figure 6.2c, so it is still highly elastic. Of course, Goodyear had no idea that rubber was a polymer. (This was discovered a hundred years later.) He did not even dream of explaining the vulcanization in the way we have just done. But his invention started the era of commercial use of new, vulcanized rubber.

The story of the discovery is very interesting in its own right. Charles Goodyear (1800–1860) was not a scientist in the modern sense of the word. His education was not very deep, and his aspirations were mainly directed toward business. Once he happened to buy a lifebuoy of india (natural) rubber. The unusual material captured the inventor's imagination. He became literally obsessed with the idea of making rubber strong and pliable. There was hardly anything Goodyear did not try! He mixed rubber with turpentine, soot, and oil. He burnt it in the oven, because the Native Americans were said to have made some progress by keeping rubber in the bright sun. There were several times when Goodyear thought he had succeeded. Then he would persuade investors to support the enterprise and immediately set up production on a really American scale. Alas, every time the rubber would start to run. The products would ooze away, sometimes giving out such a horrible smell that they had to be buried in the ground! The debts were left unpaid, Goodyear's numerous children had to live in poverty, and for a while he was even imprisoned for debt. But nothing could stop him.

To be fair, we need to say that Goodyear was not the only person working on improving india rubber. There was even a kind of "rubber fever" in the 1820s in North America. However, it was none other than Goodyear who eventually made the breakthrough.

It was entirely accidental. One day he was mixing rubber with sulphur and various other ingredients when he dropped some on top of a hot stove. The next morning the stove had cooled, and one side of the rubber lump, which was next to the sulphur, had become unrecognizable. It looked like the normal rubber we now use every day. At this stage, Goodyear did something most important, which makes him really deserve the fame, not just credit for being a lucky guy who got the answer by chance. He noticed what had happened, realized its significance, and drew the right conclusion. He saw that the recipe for success had to do both, mixing rubber with sulphur and then heating it.

This is what Goodyear wrote about his discovery:

I was encouraged in my efforts by the reflection that what is hidden and unknown and cannot be discovered by scientific research, will most likely be discovered by accident, if at all, by the man who applies himself most perseveringly to the subject, and is most observing of everything related thereto.[1]

Goodyear died almost as poor as he had been in his youth. Nevertheless, his invention became widely popular even during his lifetime. The method of vulcanization that he designed has survived till now with hardly any changes. Furthermore, many of Goodyear's ideas on how to obtain different sorts of rubber with particular features are now successfully exploited. For instance, incorporating an inert filler (such as carbon black), results in a very hard and robust rubber that is especially good for tires. (The way it works is that little particles of soot fill in the mesh of the network. This makes it harder to squash.) To obtain the opposite effect, a plasticizer (e.g., some oil that would help the particles of filler move along the network) is added. This gives rubber that is easily worn away, like that used to make erasers and the like.

Thus, since the second half of the 19th century, the rubber industry has developed very rapidly. The latex of *Hevea brasiliensis*, growing in the wild, had long remained the only raw material for the industry. However, in 1870 the English smuggled about 100,000 *Hevea* seeds from Brazil. Then young trees were cultivated from the seeds in British botanical gardens. They gave birth to vast plantations of rubber trees in the colonies, mainly in Malasia, Indonesia, and Ceylon. By the First World War, Brazil produced a negligible part of the world production of rubber.

6.4 – Synthetic Rubber

In countries that had no access to the tropical plantations of rubber trees, scientists, especially in Russia and Germany, tried to work out how to make synthetic rubber from available raw materials. The studies were a success. They found a way of making synthetic rubber from butadiene and in the 1930s production started. It worked out well; synthetic rubber was satisfactory and looked very similar to natural rubber.

All other industrial countries remained content with natural rubber—its qualities were still better overall. However, all was to change during the Second World War, when almost all the rubber plantations in Southeastern Asia were occupied by the Japanese. This encouraged the search for new methods

---

[1] Mitchell Wilson, *American Science and Invention: A Pictorial History*, Simon and Schuster, New York, 1954; page 124.

of synthesizing rubber, especially in the United States and Canada. Soon, the world production of artificial rubber had caught up with and, by the 1960s, had surpassed the production of natural rubber.

In contrast to vulcanization, this time it was all worked out scientifically. The studies on rubber synthesis kept up with the new idea that polymers were made of long molecular chains. (You may remember that H. Staudinger pioneered this theory and gave credence to it by many experiments in the 1920s and 1930s.) The endeavor with synthetic rubber not only brought in new products, it also had scientific value. Those studies confirmed that high elasticity was not a unique feature of natural rubber, but should be typical of any polymer network or gel (as long as there is no glass transition or crystallization under certain circumstances, otherwise the motion of the chains would be constrained).

6.5 — High Elasticity and Stretching of an Individual Polymer Chain

We have said that high elasticity is a common property of polymer networks. However, this sounds a bit too general. Let us zoom in and examine what it implies for particular molecules. Figure 6.2b portrays a typical polymer network. You can see a set of long molecular chains, bridged together with covalent chemical bonds. It looks like a kind of framework in three-dimension. What would you regard as an elementary "brick" of such a structure? The answer is clear from Figure 6.2b: It is a strand of the chain between two neighboring bridges (such as strand AB in Figure 6.1b). We shall introduce a new word, *subchain*, for such strands. If a polymer network is stretched, all the subchains will be stretched as well (Figure 6.2c). Therefore, the elasticity of the whole polymer is a sum of the elasticities of all the individual subchains. This is why it makes sense to explore elastic properties of a single subchain first, before looking at the whole network. Let's see what happens if a subchain is stretched by an external force $\vec{f}$ (Figure 6.3).

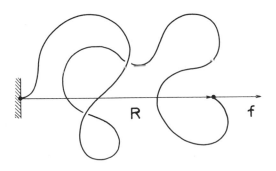

FIGURE 6.3
To keep a given end-to-end vector $\vec{R}$, we need to apply an external force $\vec{f}$.

Suppose that there is a freely jointed polymer chain made of a large number ($N \gg 1$) of elementary segments, each of length ℓ. We will neglect the volume interactions between the segments (see Section 5.3). This approximation is known as an *ideal polymer chain*; we will come back to it in Section 7.1. Imagine that one end of the chain is attached to the origin. The other end is free. Now we bring in the end-to-end vector $\vec{R}$ to describe the other end's position (Figure 6.3). Now, suppose we wanted to have a particular value of $\vec{R}$. How can we obtain it? In other words, what sort of force $\vec{f}$, on average, should we apply the dangling end to retain the chosen $\vec{R}$?

You may be a little surprised by this last question. Is a force really needed to keep $\vec{R}$ unchanged? As we learned in Section 4.7, even with no force at all $\vec{R}$ may have any possible value, including the value we wish. There is no doubt about that. However, if $\vec{f} = 0$, the dangling end will not stay at the desired point $\vec{R}$ for any length of time. It will go on fluctuating. All directions of $\vec{R}$ will be equally likely. This is why on average $\langle \vec{R} \rangle = 0$, just as you expect from the symmetry of the distribution $P_N(\vec{R})$ (5.16). Hence, to keep $\vec{R}$ fixed, you need to use a force $\vec{f}$. Because of the symmetry, $\vec{f}$ should point in the same direction as $\vec{R}$ (Figure 6.3). It is an external force, that is, a force from some external object, acting on the chain. And what about the force that the chain in its turn exerts on the external object? Newton's third law tells us that it should point in the opposite direction, toward the origin of coordinates. So it is a restoring, elastic force. Thus, we have naturally come to the conclusion that a polymer chain resists being stretched. In other words, it exhibits elasticity.

Our arguments may not satisfy you completely. You may be convinced that a polymer chain should behave as an elastic body. But what is the physics of such elasticity? To answer this, let's think of an ordinary solid crystal. What makes it elastic? When a crystal is stretched, the atoms are pulled further apart (Figure 6.4). Thus, the elastic force in this case would be a result of many interatomic interactions. Sometimes it helps to describe the same thing in terms of energy. The undeformed crystal is in equilibrium; that is, the potential energy of interatomic interactions is a minimum (Figure 6.4). An external deforming force pulls the atoms up the slope from the bottom of the potential well. For example, suppose an external force f causes an elongation of the crystal, Δx. The work it does, $f\Delta x$, is used to increase the internal potential energy ΔU of the crystal, which is the total energy of interatomic interactions: $f\Delta x = \Delta U$, or

$$f = \frac{\Delta U}{\Delta x}. \qquad (6.1)$$

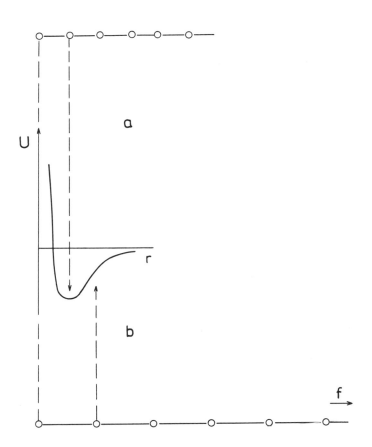

FIGURE 6.4
An illustration of the elasticity of a crystal. (a) Initially, the distance between the atoms corresponds to the minimum of the interactional potential energy $U(r)$. (b) When the sample is stretched, the potential energy is increased.

Equation (6.1) gives you a recipe for finding the stretching force f (and the elastic force of the crystal, which is opposite to it). All you need to know is how the crystal is constructed. Then you should be able to work out its internal energy ΔU.

Now let's go back to a polymer chain. This time, we are not talking about a stretched array of atoms, as in the case of a crystal. What we really see when a polymer chain is stretched is an increase in the end-to-end distance (Figure 6.3). The only way this can happen is, of course, if some wiggled bits of the chain straighten up and disentangle. We may find it easier to follow this if we return to a freely jointed polymer (Figure 2.5). Moreover, think of the chain as ideal. (Are you now getting used to this approximation? It assumes that the only interaction between the neighboring monomers comes from their being joined together into a chain. All other monomers do not interact at all, just like the molecules in an ideal gas.)

FIGURE 6.5
An ideal gas in a vessel
with a piston.

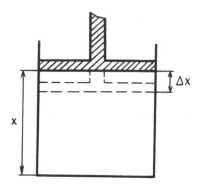

Now suppose that we apply a force $\vec{f}$ and change the end-to-end vector $\vec{R}$ by some value $\Delta\vec{R}$. [2] Hence, we did an amount of work $\vec{f}\Delta\vec{R}$. Where did the energy go? Previously, when we looked at crystals, we managed to find the answer quite easily. Unfortunately, the way we did it would not work in this case. In an ideal system, interactional potential energy is zero, both before and after the deformation. Thus, equation (6.1) is of no use. As for kinetic energy, it is determined by the temperature. If the temperature is constant during the deformation, the kinetic energy will not change either.

Before giving up, let's think of another analogy. Strange as it may seem, help comes from an ideal gas. In a sense, an ideal gas has "elasticity the other way round." Suppose you wanted to hold the gas under a piston (Figure 6.5) in a vessel of a certain volume (which is just like holding a polymer chain to retain a nonzero $\vec{R}$). You would have to apply a squeezing force $f = pA$, where p is the gas pressure and A is the area of the piston. To reduce the volume, you would always need to do some work. Nevertheless, both the potential and kinetic energy of the gas molecules stay the same, given that the compression is isothermal. So where does the work go? (As you see, we end up with the same question once again.)

Of course, in the long run, all the work transforms into heat, which is dissipated in the surroundings. How do we know? Well, if there were no surrounding medium to take the heat away, both a gas when compressed and a polymer when stretched would get warmer (see Section 6.11).

Is there any hope we can still manage with simple energy arguments? Or do we need to tackle the problem by means of mechanics, tracing the molecules?

---

[2] Both before and after the strain, the chain ends dangle randomly. Having said that, in a nondeformed coil the average end-to-end vector is zero, $\langle\vec{R}\rangle = 0$ (as we have seen in Chapter 5). The strain serves to pull it out to a nonzero value, $\langle\vec{R}\rangle = \Delta\vec{R} \neq 0$.

It would not be hard for an ideal gas. Its pressure is just an overall result of all the individual collisions between the molecules and the piston. This concept immediately leads to Mendeleev–Clapeyron equation of state. However, there is no such simple picture for a polymer. Even in the case of an ideal chain, the motion of segments is extremely complicated, due to knots and entanglements.

But let's think. We know that the surrounding medium plays no other role but to maintain the constant temperature T. Our proof of polymer elasticity was based on a very general idea. Indeed, we showed that when a polymer chain is stretched, it is pushed from a more probable to a less probable state. Hang on a minute! Is there perhaps some universal way of finding the energy cost of such a transition between two states with different probabilities at a constant temperature, without getting bogged down in the mechanics of molecular collisions?

There is indeed a very general rule, known as the Boltzmann principle. It states the following. Suppose there are W ways in which molecules can occupy a certain state. (In our case, this number is proportional to the probability $P(\vec{R})$ (see equation (4.14).) Then we need to find the quantity

$$S = k \ln W, \tag{6.2}$$

where k is Boltzmann's constant. The energy equivalent of probability we are seeking is the change in the value;

$$U_{eff} = -TS, \tag{6.3}$$

where T is the thermodynamic temperature. In the case of a polymer chain, according to (4.14),

$$S(\vec{R}) = -k\left(\frac{3\vec{R}^2}{2N\ell^2}\right) + \text{const}, \tag{6.4}$$

or

$$U_{eff}(\vec{R}) = -kT\left(\frac{3\vec{R}^2}{2N\ell^2}\right) + \text{const}, \tag{6.5}$$

where const is a quantity independent of $\vec{R}$. Using equation (6.5), we can easily find the elastic force from (6.1). We shall do so a little later, in Section 6.9. Thus, the Boltzmann equation $S = k \ln W$ has rescued us when we had nearly lost hope. It is not surprising that this formula was engraved on the tombstone of the author,

Ludwig Boltzmann (1844–1906)! But what does it mean, and where does it come from? It is an interesting question in its own right. We shall devote the next two sections to it, and then come back to the discussion of high elasticity.

6.6 — Entropy

Rapidly developing science gives rise to a new vocabulary. This gives us an excuse to reflect on how human languages evolve. It is fascinating to be able to trace this process, spanning from the dawn of mankind to the modern day. For example, officers of the Russian army, after entering Paris in 1815, used to spur on French waiters in Russian. The Russian word *bystro* (meaning "quickly") soon became absorbed into French. Much more recently, we witnessed how the English language acquired such strangers as *sputnik* and *perestroika*, whereas the Russians borrowed words like *computer*. Novel words usually enrich the language, as they represent new things and ideas. As soon as the words are borrowed, their meanings may become slightly different from the original, since they are no longer associated with the context in which they first appeared. (For example, in Russian, computer no longer means "something that calculates." Likewise, the word *sputnik* in English does not mean satellite in general (which would have been the correct translation), but rather refers to early Russian satellites, and so evokes memories of that period of time and the mood of that time).

Some newly invented or borrowed scientific words may sound pretty horrible (like *uniformitarianism* or *turbidimetry*). Such words have a very narrow use and clutter up the language. (We honestly think that their originators must have lacked a sense of moderation!) Nevertheless, there have been some really valuable contributions to the world's languages over the last hundred years. The word *entropy* is among them, and it certainly deserves a place near the top of the list.

Together with energy, time, and so on, entropy is one of the most crucial concepts of physics and of science in general. Since the idea of entropy appeared, it has always been surrounded by a halo of mystery. Even in the 20th century, the well-known German physical chemist W. Ostwald put it this way: "Energy is the queen of the world, and entropy is her shadow!" Such an attitude is not without reason. How do people hear about entropy the first time? Quite often it gets mentioned when the most global and tantalizing problems, such as the origin of life or the future of the Universe, are discussed.

Perhaps this explains why there is usually no room for entropy in the school curriculum. However, it is quite a straightforward thing. To get to know it in the first instance, you do not need to dive into obscure philosophical matters. Moreover, it is hard to manage without entropy, if you are aiming to describe atomic properties of matter. It would be a bit like trying to explain the rules of football without mentioning the ball! This is especially true for polymers. Now you understand why we need to digress from the main theme and talk about entropy in more detail.

Let's think of energy, for a start. How would you define it? Of course, you can split it up into various forms—potential energy, kinetic energy, and so on—and describe each separately. However, the real meaning of energy is revealed by the conservation law. Consider a complex system. Suppose we know that somewhere in this system a certain form of energy has decreased. This means that the energy of the other parts must have increased (given that the system is isolated). Thus, we are able to draw the right conclusion straightaway. We don't even need to know anything about the way the system functions or what it is made of.

Now back to entropy. Equation (6.2) can be regarded as its definition. As we have already said, entropy is the energy equivalent of probability. In other words, if you look at how much the value $(-TS)$ has changed, it will tell you exactly how much work has been done to transfer the system from a more probable to a less probable state. In this case yet again, just like with energy, you need not worry about the details—what did the work (a piston, an electric field, etc.), how the molecules collided with the object doing the work, and so forth.

What exactly does the Boltzmann principle (6.2) mean? Its main idea is that the quantity $U_{eff} = -TS$ defined by (6.3) and (6.2) can be regarded as a sort of potential energy. Indeed, if the system is left to itself, it is most likely to drop down into the most likely state (sorry for this tautology!). According to (6.2) and (6.3), this would mean an increase in entropy and hence a decrease in U_{eff}, which is just what the principle of minimum potential energy predicts.

Figure 6.6 sketches the function $U_{eff}(\vec{R})$ for an ideal polymer chain, in accordance with (6.5). The graph has the shape of a potential well. However, you cannot say that "sitting" at the bottom of the well corresponds to the equilibrium. We are talking about nonzero temperatures here. Suppose you have a little ball at temperature T, and you put it into a proper potential well (not U_{eff}, but merely U). What will it do? It will go jittering around the equilibrium position, in a random Brownian way. The typical size of the swings will be such that the potential energy increases by about kT. (By the way, this is just how physicists estimate the amplitudes of thermal oscillations of atoms in a crystal.) A similar thing can be said about U_{eff}. As you can see from equation (6.5), the condition $U_{eff}(\vec{R}) - U_{eff}(0) = kT$ leads to the requirement that the distance $|\vec{R}| = N^{1/2}\ell$.

FIGURE 6.6
The dependence of the
effective potential
energy $U_{eff} = -TS$ of a
polymer on the
end-to-end distance R.
This picture shows how
the amplitude of the
fluctuations in R can be
found.

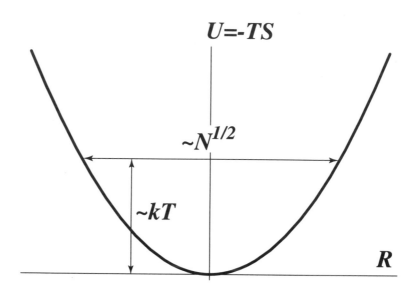

This is exactly what we want—the most probable end-to-end distance for a single polymer coil (see Section 5.7).

Now the Boltzmann equation has become a little clearer, because we have sorted out U_{eff} and agreed that it is something like potential energy. Yet it is so tempting to actually try to derive the equation! Thus we will do it for an ideal gas, in the next section.

6.7 — Entropic Elasticity of an Ideal Gas

Assume there is an ideal gas in a vessel with a piston. The Mendeleev–Clapeyron equation gives us the pressure of the gas, p:

$$pV = NkT, \tag{6.6}$$

where N is the total number of molecules and V is the volume of the vessel. The force acting on the piston will be

$$f = pA = \frac{NkTA}{V}, \tag{6.7}$$

where A is the surface area of the piston.

Suppose we have pushed the piston down by Δx and slightly compressed the gas. The volume of the vessel has obviously decreased by $\Delta V = A\Delta x$ (Figure 6.5). If Δx is very small, we can neglect the tiny variations of the force and pressure while the piston is moving. Hence the work done will simply be given by

$$f\Delta x = pA\Delta x = p\Delta V = \frac{NkT\Delta V}{V}. \tag{6.8}$$

At this stage, it would help if we remembered that $\Delta V/V$ behaves like $\Delta(\ln V)$, in the limit of the small ΔV. Indeed, from calculus,

$$\Delta(\ln V) \approx \frac{\partial(\ln V)}{\partial V}\Delta V = \frac{\Delta V}{V}. \tag{6.9}$$

Since N is constant,

$$\frac{N\Delta V}{V} = N\Delta(\ln V) = \Delta(N \ln V) = \Delta(\ln V^N). \tag{6.10}$$

Hence, we get

$$f\Delta x = kT\Delta \ln V^N, \tag{6.11}$$

or

$$f = -\frac{\Delta U_{eff}}{\Delta x}, \tag{6.12}$$

where

$$\Delta U_{eff} = -kT\Delta \ln V^N. \tag{6.13}$$

So we have the force in the form (6.12), which looks similar to (6.1). Now we need to specify what U_{eff} is. We are going to link it with the number of ways in which a molecule can be positioned in the vessel.

Clearly, a gas cannot decrease in volume of its own accord. Yet it can expand with no extra help. This inequality of rights has to do with the difference in probabilities. Apparently, a ramified state of gas is more probable than a denser state. The reason is that it can be realized in more different ways. But how do we know? Can we really count all the possible ways for each state? The answer is yes, and the simplest procedure is the following. Let's divide the whole volume of the gas into little cubic cells, of volume a^3 each (Figure 6.7). To make it simpler, let's

FIGURE 6.7
Counting the number
of states of an ideal
gas.

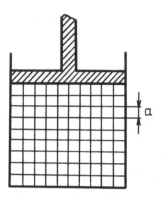

assume that each molecule can only be located in the centers of the cells. Hence, we are bringing in a discrete distribution of the molecules' positions, instead of a continuous one. There will be V/a^3 different ways of accommodating each molecule in the volume V. Suppose there are N molecules all together. Then you will be able to arrange them in $(V/a^3)^N$ ways in the volume V. This is because the molecules of an ideal gas do not interact with each other; so you can spread them around the cells totally independently. If the piston was originally at a height x_1, the initial volume was $V_1 = Ax_1$, whereas the final volume is $V_2 = Ax_2 = A(x_1 - \Delta x)$. The number of ways, W_1 and W_2, in which the two states can be realized is given by

$$W_1 = \left(\frac{V_1}{a^3}\right)^N \qquad W_2 = \left(\frac{V_2}{a^3}\right)^N. \qquad (6.14)$$

Obviously, $W_1 > W_2$ (because $V_1 > V_2$), so indeed the final, more compressed state is the less probable one. However, you may feel a bit suspicious about the method of calculation. We jumped rather carelessly from the continuous set of the molecules' positions to a discrete one. In fact, it is quite all right to do so. You will see a little later that the size of a cell a will cancel out from all the final formulas for physical quantities. This means that it does not matter after all how you break up the volume and count the states.

How much do the probabilities W_1 and W_2 really differ? Let's look at the ratio W_1/W_2. Equation (6.5) leads to

$$\frac{W_1}{W_2} = \left(\frac{V_1}{V_2}\right)^N. \qquad (6.15)$$

The ratio V_1/V_2 is greater than one; the index N is a large number, much greater than one. Therefore, $W_1/W_2 \gg 1$. Let us make a simple estimate. Say, for example,

$V_1/V_2 = 1.01$ and $N = 6 \cdot 10^{23}$ (this is roughly the number of molecules in one mole of a gas). Then

$$\frac{W_1}{W_2} = (1.01)^{6 \cdot 10^{23}}, \tag{6.16}$$

which is an incredibly enormous number; it is even greater than the number of atoms in the whole Universe! Thus, according to (6.5), the final state is incomparably less likely than the initial one.

It seems obvious that spontaneous evolution of a system can only go in one direction, from a less probable to a more probable state (especially when there is such an immense gap between the two probabilities). Now you understand why a gas can spontaneously expand (at a constant temperature), but is unable to shrink of its own accord. To compress the gas, you have to push the piston with some force. The counter reaction against your effort is the elastic force of the ideal gas.

Using (6.15), we can rewrite expression (6.11) in terms of W:

$$\Delta \left(\ln V^N \right) = \ln V_1^N - \ln V_2^N = \ln \left(\frac{V_1}{V_2} \right)^N = \ln \left(\frac{W_1}{W_2} \right) = \Delta \left(\ln W \right). \tag{6.17}$$

Comparing this with (6.13), we can conclude that, for an ideal gas,

$$U_{eff} = kT \ln W, \tag{6.18}$$

which is the same as the Boltzmann equation (6.2).

6.8 Free Energy

There is one more tricky bit that we need to sort out. What if we have a complex structure where both the potential energy U and the probability W of the state (or the entropy S as in (6.2)) are changing at the same time? In fact, this is exactly what happens in practice, in real physical systems including polymers.

The answer is obvious. During any isothermal process, an external force has to do work on both things at the same time, that is, altering the energy U as well as the number of states W. The work done is determined by the change in the total $U + U_{eff}$. This quantity is known as the free energy F and is normally written as

$$F = U + U_{eff} = U - TS. \tag{6.19}$$

This leads us right to the very basic principle of minimum free energy. Any system, left on its own, always behaves in such a manner that its free energy goes down. The minimum of the free energy corresponds to the equilibrium state. However, the equilibrium is only defined in a statistical sense—the system never stops its random thermal jittering around the equilibrium position. We say that it fluctuates.

The two terms in the free energy (6.19) are often known as the "energy part" and the "entropy part." Using this, we can make some interesting generalizations. An ideal gas and an ideal polymer both appear to have a zero energy part of the free energy. On the other hand, an ideal crystal has a zero entropy part.

Later on, we shall have a few more chances to explore free energies of polymers. But for now, let's answer the question you might have had for some time: Why "free"? What a strange name! In fact, the concept of free energy (as well as of entropy) belongs to thermodynamics, which is known as "the child of the age of steam." It was once a very applied area, concentrating on the problems of heat-engine design. Even now, if you look in the wrong textbook, you might get the impression that thermodynamics is a strange, out-of-date study of steam engines. This is certainly far from true. Thermodynamics is a probably unique example of a science that originated from rather narrow practical problems and gradually formed into a very general field of knowledge, spanning from cosmology to biology.

It is most amazing that scientists such as Carnot and Clausius, who laid the foundations of thermodynamics, still believed in a very naive caloric theory of heat, which held that heat is a form of fluid. The main practical question they faced was the following. Suppose there is some hot steam. How much of its energy can it give away to produce useful work? (Presumably, it cannot lose all its heat!) In other words, what fraction of the energy is free to be converted into work? The answer is: the free energy (6.19)! Hence the name.

6.9 — Entropic Elasticity of a Polymer Chain

Let's now go back to the high elasticity. As we have just seen, the internal energy of an ideal polymer does not change $\Delta U = 0$. So there is no energy contribution to the elasticity; the elasticity is explained in terms of entropy alone. Indeed, when a chain is stretched, we move from a more probable state (realized in more different ways) to a less probable one (realized in fewer ways). The coil starts getting disentangled and loses some freedom. In the extreme case, a chain stretched out in a straight line has no freedom at all ($W = 1$, $S = 0$).

We have already found the entropy of an ideal chain (6.4). Now we can use formula (6.1) to find the elasticity:

$$f = -T\frac{\partial S}{\partial R} = \left(\frac{3kT}{N\ell^2}\right)R. \tag{6.20}$$

The vectors $\vec{f}$ and $\vec{R}$ are parallel (as you can see in Figure 6.3). This is why we can rewrite (6.20) in the vector form:

$$\vec{f} = \left(\frac{3kT}{N\ell^2}\right)\vec{R}. \tag{6.21}$$

Thus, the force $\vec{f}$ has turned out to be proportional to the "displacement" $\vec{R}$. We can say that an ideal chain obeys the well-known Hooke's law. However, perhaps we need to be a bit more cautious. Compare (6.20) with an ordinary form of the law (3.1). The main discrepancy is that the average value of $\vec{R}$ in a nondeformed chain equals zero. Therefore, we cannot bring in anything like the relative deformation $\Delta\ell/\ell$, which appears in the usual form of Hooke's law.

Still, we could think of an "elastic modulus" of a polymer chain. As usual, it would be the ratio of the force $\vec{f}$ to the deformation $\vec{R}$. According to (6.20), it happens to be $3kT/N\ell^2$. First, notice that it is proportional to $1/N$, which makes it a very small quantity if the chains are fairly long. This means that polymer chains are very susceptible to external forces; this is exactly what accounts for the high elasticity of rubber and other similar polymers. The second thing we notice is that the elastic modulus is proportional to the temperature T. This is because the elastic forces are due to entropy, as you can see from (6.3).

6.10 – Entropic Elasticity of a Polymer Network

We have explored what happens when an individual polymer chain is stretched. This was not just an exercise. We have shown that the elasticity of a network is built up from the elasticities of all the subchains (Figure 6.2), so we can make use of what we have found. There is one tricky question though. Let us imagine a highly elastic solid body, say, a rubber ball. The macromolecules are rather closely packed in it and interact strongly with each other. So can we really treat each subchain as an ideal polymer, with no volume interactions at all?

The answer is that we can. Of course, in such a dense structure, the thermal motion of the molecules will be nothing like that of an ideal single chain. Atomic groups within one monomer will oscillate and rotate in a totally different fashion. However, the density of the surroundings will make no difference to the entangled shape of the macromolecules (i.e., the size of a coil will still be proportional to the square root of the chain length).

The Gaussian distribution (5.16) will not be affected either. In general, the large-scale properties of chains are the same for both ideal and highly elastic polymers. This idea was voiced clearly for the first time by P. Flory in 1949; thus it is often called the Flory theorem. You can explain it qualitatively in this way. In a uniform, amorphous substance all the conformations of a certain chain are equally likely (in the sense that they correspond to the same energy of interaction with the other chains). This is because the surroundings of each unit are roughly the same. But this is the only assumption we actually made when deriving the elasticity of an ideal polymer.

Now, that we are convinced we are on the right track, let's investigate the stretching of a polymer network (6.2). We shall treat it as a set of ideal subchains. Suppose each subchain consists of N freely jointed segments, each of length ℓ. (To make it simpler, we neglect the polydispersity of the polymer.) When the network is stretched, all the subchains are also stretched on average. Their entropy (6.4) decreases (as the end-to-end distance R grows). This causes an "entropic" elastic force. It does not explain the high elasticity yet. The high elasticity is the capability of bearing huge reversible strains at rather moderate stresses. It occurs because the "elastic modulus" of each chain is fairly small (see (6.20)).

Imagine a polymer network in the shape of rectangular parallelepiped. Let's draw the x-, y-, and z-axes along its sides. Suppose we have elongated the network by factors λ_x, λ_y, and λ_z along these axes (respectively). Then, if the initial length of the network along the x-axis was a_{0x}, it will now be $\lambda_x a_{0x}$, and so on. Now we need to make some assumption about how the network is deformed. The simplest is to assume what is called affinity (where the cross-links and the whole network deform in the same way). Suppose the end-to-end distance of a certain subchain was initially $\vec{R}_0$ (i.e., its components were R_{0x}, R_{0y}, and R_{0z}). After the deformation, the vector becomes $\vec{R}$ and its components are $R_{0x}\lambda_x$, $R_{0y}\lambda_y$, and $R_{0z}\lambda_z$. According to (6.4), the change in entropy of the subchain is

$$\Delta S(\vec{R}) = S(\vec{R}) - S(\vec{R}_0)$$

$$= -\frac{3k}{2N\ell^2}\left[\left(R_x^2 - R_{0x}^2\right) + \left(R_y^2 - R_{0y}^2\right) + \left(R_z^2 - R_{0z}^2\right)\right]$$

$$= -\frac{3k}{2N\ell^2}\left[\left(\lambda_x^2 - 1\right)R_{0x}^2 + \left(\lambda_y^2 - 1\right)R_{0y}^2 + \left(\lambda_z^2 - 1\right)R_{0z}^2\right]. \quad (6.22)$$

To find the total change in the entropy of the whole network, we have to sum equations like (6.22) over all the subchains. In other words, we can average over $\vec{R}_0$ and multiply by the number of subchains, νV, in the network. (Here V is the volume of the sample, and ν is the concentration of subchains per unit volume.)

$$\Delta S = -\frac{3k\nu V}{2N\ell^2}\left[\left(\lambda_x^2 - 1\right)\langle R_{0x}^2\rangle + \left(\lambda_y^2 - 1\right)\langle R_{0y}^2\rangle + \left(\lambda_z^2 - 1\right)\langle R_{0z}^2\rangle\right].$$
(6.23)

Now we can take into account that

$$\langle \vec{R}_0^2\rangle = \langle R_{0x}^2\rangle + \langle R_{0y}^2\rangle + \langle R_{0z}^2\rangle = N\ell^2$$
(6.24)

(see (5.11)). We also know that all the three directions (x, y, and z) have equal rights, therefore $\langle R_{0x}^2\rangle = \langle R_{0y}^2\rangle = \langle R_{0z}^2\rangle = N\ell^2/3$. So we finally get

$$\Delta S = -\frac{k\nu V \left(\lambda_x^2 + \lambda_y^2 + \lambda_z^2 - 3\right)}{2}.$$
(6.25)

It is interesting that the answer does not depend on the parameters N and ℓ that describe an individual subchain. This indicates that equation (6.25) is universal. It works whatever the particular structure of the subchains (regardless of whether they are freely jointed or wormlike), for whatever contour lengths and Kuhn lengths, and so on. If we glance again at our calculations, we can see that basically all we needed to draw the main conclusion (6.25) was just to regard the subchains as ideal.

We can use (6.25) to find the stress caused by the "entropic" elasticity for all sorts of deformations. Obviously, one of the most important types of deformation is the uni-axial elongation (or compression). Let's see what we can get out of (6.25) in this case. Suppose we have elongated the sample by the factor of λ along the x-axis, that is, $\lambda_x = \lambda$. The size of the network along the y- and z-coordinates may change freely. Can we find the relative deformations λ_y and λ_z in this case?

Remember that we are talking about a polymer in a highly elastic state. It seems a sensible assumption that its volume has not changed under the strain. Then, both the y-size and the z-size of the sample ought to have shrunk by a factor of $\lambda^{-1/2}$, that is, $\lambda_y = \lambda_z = \lambda^{-1/2}$. Thus the total volume after the deformation would be

$$V = \lambda_x a_{0x}\lambda_y a_{0y}\lambda_z a_{0z} = \lambda_x\lambda_y\lambda_z V_0 = V_0 = \text{const.}$$
(6.26)

How can we justify, physically, that the volume has to be constant? A highly elastic polymer is usually a sort of fluid (a polymer melt). Its chains are linked with chemical bonds. So, if squashed in all directions, such a polymer is bound to behave as an ordinary liquid. In particular, a 1% change in volume can only be achieved with a pressure of roughly 100 atm $\sim 10^7$ Pa). At the same time, the elastic modulus of such polymer is fairly small. Therefore, the sample can be stretched a few times its length with a much smaller stress ($\sim 10^5$ or 10^6 Pa). So it is only natural to assume that the volume does not change under such low stresses.

From this point of view, elastic polymers are different from ordinary solid crystals and glasses, which change their volume just because their length changes. At a molecular level, this difference is not surprising. When crystals are elongated, their atoms are pulled farther apart. Meanwhile, polymers increase their length by merely disentangling and stretching out their wiggly subchains; this way the distances between the atoms are kept unchanged.

If we substitute $\lambda_x = \lambda$ and $\lambda_y = \lambda_z = \lambda^{-1/2}$ into equation (6.25), we obtain

$$\Delta S = -k\nu V \frac{(\lambda^2 + 2)/(\lambda - 3)}{2}. \tag{6.27}$$

There is no problem in finding the elongating force here, using a formula similar to (6.12):

$$f = -T\frac{\Delta S}{\Delta a_x} = -\frac{T}{a_{0x}}\frac{\Delta S}{\Delta\lambda} = -\frac{T}{a_{0x}}\frac{\partial S}{\partial\lambda}. \tag{6.28}$$

More often we are not interested in the force as such, but rather in the stress, that is, the force per unit area of the initial cross section:

$$\sigma = \frac{f}{a_{0y}a_{0z}} = -\frac{T\partial S/\partial\lambda}{a_{0x}a_{0y}a_{0z}} = -\frac{T\partial S/\partial\lambda}{V}. \tag{6.29}$$

Therefore, let us rewrite our answer in terms of σ:

$$\sigma = kT\nu\left(\lambda - \frac{1}{\lambda^2}\right). \tag{6.30}$$

The result (6.30) is a major one in the classical theory for the high elasticity of polymer networks. If the elongation is small (i.e., λ is close to one), equation (6.30) can be used to estimate Young's modulus of a polymer network (see (3.1)).

Indeed, in the limit of $\lambda \approx 1$:

$$\lambda - \frac{1}{\lambda^2} = (\lambda - 1) + (1 - \lambda^{-1})(1 + \lambda^{-1}) \approx \lambda - 1 + \frac{2(\lambda - 1)}{\lambda} \approx 3(\lambda - 1). \quad (6.31)$$

Meanwhile, the value $\lambda - 1 \equiv (a_x - a_{0x})/a_{0x}$ is just the relative elongation. In other words, it plays the same role as the parameter $\Delta \ell / \ell$ in equation (3.1). Comparing (3.1), (6.30), and (6.31), we end up with Young's modulus:

$$E = 3kT\nu. \quad (6.32)$$

Thus, E turns out to be the same as the pressure of an ideal gas whose molecular concentration is 3ν (i.e., three times the concentration of the cross-links). This means that the more cross-links there are in a highly elastic sample, the less elastic it is. Therefore, the value of E does not indicate a specific polymer. It varies dramatically depending on the density of the cross-links.

However, (6.30) can be used not only to find Young's modulus, it also describes the nonlinear elasticity, which takes up quite a lot of room on the stress versus strain curve. (In Figure 6.1, it spans from point A, where the linearity disappears, up to point B, where the reversibility is lost.) What is more, equation (6.30) is just as good for uni-axial compression. You only need to bear in mind that λ will be less than one in this case. Another warning is that when compressed along the x-axis, the sample will automatically stretch in both the y- and z-directions, accordingly. Even more complex deformations, such as two-dimensional elongation, torsion, shear, and so on, are covered by the general relationship (6.25). So you could derive equations similar to (6.30), revealing nonlinear behavior of the stress. However, we shall have to leave this out of our discussion.

How good is the theoretical prediction (6.30) for $\sigma(\lambda)$ compared with experiments? Figure 6.8 brings together both a typical experimental curve and the theory. You can see that up to $\lambda = 5$ the agreement is more or less tolerable. Then, for $\lambda > 5$, the discrepancy grows more and more. This is not surprising. Expression (5.16) for $P_N(\vec{R})$ ceases to work for long end-to-end distances R (or, equivalently, large elongations). Why? Because it does not take into account that there is a limit to how much the chains can actually be stretched. Namely, the distance R can never exceed the total contour length $N\ell$. This is why equation (6.19) and all the consequent results will not hold for this case.

Let's look at the range of moderate elongations: $1.2 < \lambda < 5$. For most polymer networks, typical discrepancies between the theoretical and experimental $\sigma(\lambda)$ are not that high (about 20% or so), but they tend to be systematic

FIGURE 6.8
The dependence of
$\sigma(\lambda)$ for highly elastic
polymer networks. The
solid line is the theory
(6.30); the dots show a
typical experimental
curve.

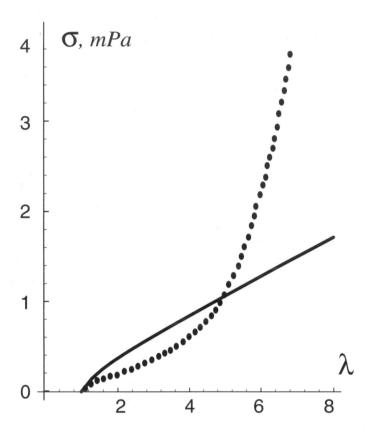

(Figure 6.8). These are explained by the so-called topological constraints to the subchains' conformations (see Section 2.6).

Despite all the disagreements with experiment, this approach to the high elasticity has proved quite successful, especially because it is universal. All the predictions are satisfactorily accurate, whether it is absolute values of Young's moduli, or their temperature dependences, or the shape of the nonlinear stress versus strain curve. There are not many other examples in the physics of disordered solids and liquids where such simple arguments have helped to understand so much. The reason is rather obvious. The way the chains are entangled and rolled up into coils is not described by the short-scale chemical structure or interactions of individual atomic groups. It is determined by the very fact that the monomers are grouped into chains. So it is a long-scale feature. Soon we shall have a chance to discover that some other most interesting and peculiar properties of polymers have the same sort of origin.

6.11 The Guch–Joule Effect and Thermal Phenomena During the Deformation of Rubber

Until now, all we have deduced from (6.30) was the dependence $\sigma(\lambda)$ at constant temperature. However, it also contains the dependence on T. So, to conclude this chapter, we will analyze how the elasticity of polymer networks is affected by the temperature.

Suppose we hang a weight on a rubber string. The string will become elongated. Now let's increase the temperature. According to (6.30), as long as $\sigma = \text{const}$, an increase in the temperature should lead to a decrease in λ. Thus, a stretched rubber string, in contrast to most nonpolymer materials, contracts on heating! This strange behavior was discovered by Guch as early as 1805 when he experimented with strips of natural rubber. Half a century later, Joule carried out a careful set of measurements to confirm Guch's result. Thus this phenomenon is usually known as the Guch–Joule effect.

This effect demonstrates, perhaps in the most dramatic way, that the high elasticity of rubber and other polymers is related to entropy. Indeed, as the temperature goes up, all sorts of interactions start to lose their importance. This is because the characteristic energy of such interactions, ε, becomes much less than kT (i.e., $\varepsilon/kT \simeq 1$). Meanwhile, the entropy contribution gains more and more significance. (According to (6.4), the entropic elasticity is proportional to the temperature.) Therefore, the fact that the "bouncing" reaction is enhanced by heating suggests that entropy is to blame for the high elasticity of rubber.

What happens when the system cools down again? The sample becomes partially crystallized, and the Guch–Joule effect is replaced by the opposite. Now the partially crystallized rubber would expand with heating. This is yet another sign that the high elasticity of polymers has a very special nature.

A common but very impressive way to demonstrate the Guch–Joule effect in a lecture setting is the following. (This can be reproduced easily in an ordinary school laboratory.) Using a bicycle wheel, replace the spokes by elastic strings made of soft rubber (Figure 6.9). Preferably, the strings should be stretched to about three times their original length. Now fix the axle in a horizontal position, so that the wheel can rotate in the vertical plane with little friction. Place an electric heater pointing at a certain part of the wheel. (You can also use a powerful electric light bulb for this purpose, shining on some section of the wheel.) The heat makes the rubber strings shrink, and the center of mass shifts. As a result, the warmed sections of the wheel move up, and other sections take their place at

FIGURE 6.9
An experiment to
demonstrate the
Guch–Joule effect.

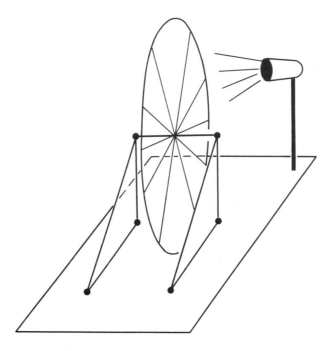

the spot being heated. They are warmed up in their turn, and so on. As you might have guessed, the wheel starts rotating at a constant angular speed.

Here is an even simpler experiment. Hang a weight on a rubber band, and start warming up the band using either a light bulb or some other sort of electric heater. Alternatively, you may place the whole thing in water (say, in a bucket) and heat it up. When heated, the rubber will pull the weight up and upon cooling it will let it down again.

In the Guch–Joule phenomenon, the change in size of an elastic string is a consequence of the temperature variation. It turns out that it works the other way around, too: The temperature of a string will change if its size is changed, for example, if it is rapidly squashed or elongated. You can check it yourself. Stretch a rubber band quickly, then touch it with your lips to feel how it warms up! In contrast, when compressed, it will gradually cool down. This is totally opposite to what happens with an ideal gas, which gets warmer when rapidly compressed and cools when expanded rapidly.

In spite of the contrast, the two types of behavior have the same cause. Say a system is quickly extended (expanded) or compressed. Obviously, there will be no time for heat to be exchanged with the surroundings. In other words, all such processes can be regarded as adiabatic. So why should an ideal gas warm up when it is adiabatically squashed? The answer is this: The squashing requires some work to be done by external forces. Where does this work go? Since no heat is leaking out, all the work is transformed into the internal energy of the gas. Therefore, the temperature rises. The same happens when a piece of rubber is adiabatically (rapidly) stretched. Again, external forces do the work, which is all used to increase the rubber's internal energy (and the temperature). In contrast, when either rubber is compressed or an ideal gas is expanding, the work has to be done by the system itself. As no extra heat comes from the outside, some of the system's own internal energy has to be used, and so it cools down.

7

The Problem of
Excluded Volume

7.1 — Linear Memory and Volume Interactions

What are the chances that one or another theoretical study will be a success? As history shows, it greatly depends on whether theorists can think of a nice, manageable model idealizing the real world. Of course, there are no ideally simple systems in nature. However, we can use our imagination and invent an ideal gas (whose molecules do not interact), an ideal liquid (where there is no internal friction at all), an ideal crystal (with perfectly regular atomic structure), and so on. As a matter of fact, you can say that all these models are ideal indeed, meaning that they are the best for physicists. This is because they are the simplest, most basic ones. So one has to master them first, before moving any farther in either statistical mechanics, or hydrodynamics, or solid state physics, or whatever.

How crude are the results we might get from such "ideal" models? Are there some cases where they work well and some where they fail? There is a special trick that often helps us to decide. It involves finding some dimensionless parameters, either large or small, that describe the system. For example, a gas can be characterized by the fraction of its volume taken up by the molecules. If this

109

parameter is much less than one, the gas can be treated as ideal. An ideal liquid approximation can be used if the energy losses due to internal friction are much less than the characteristic kinetic energy of the liquid itself. Similarly, a crystal is nearly ideal if typical displacements of the atoms from the equilibrium positions are much smaller than the interatomic distances.

What sort of large or small dimensionless parameters can describe a polymer? One of them we have actually used already, and not just once. It is the large number of monomer units ($N \gg 1$) in a chain. We have shown that a huge N can account for many things. It explains, for example, the low concentration of monomers in a coil (see (4.12)), the existence of semidilute solutions (Figure 3.7c), and the high elasticity of polymers.

Another special polymer parameter comes from the hierarchy of interactions. The energy E_1 of a covalent bond between two neighboring monomers in a chain is normally about 5 eV $\approx 0.8 \cdot 10^{-18}$ J. This is much higher than the typical energy E_2 of any other interactions (say, between the polymer and the solvent, or between monomers that are not nearest neighbors along the chain, etc.) Roughly $E_2 \sim 0.1$ eV $\approx 1.6 \cdot 10^{-20}$ J. Therefore, the ratio $E_2/E_1 \simeq 1$ is just the type of small parameter we were seeking. It allows us to introduce an ideal polymer chain approximation.

Indeed, let's see what happens near room temperature ($kT \sim 3 \cdot 10^{-2}$ eV $\sim 0.5 \cdot 10^{-20}$ J). This region is the most interesting one as far as polymer properties are concerned. Covalent bonds cannot be broken, either due to thermal fluctuations (since $E_1/kT \gg 1$) or because of interactions. This means that the sequence of units is "cemented" into the chain by the high energies of the lengthwise covalent bonds. Each unit "remembers" its own number, which it acquired when the chain was formed. To put it briefly, a polymer chain has a fixed linear memory.

Having sorted out the covalent bonds between the neighbors, we can now concentrate on all the other interactions. These are usually referred to as "excluded volume interactions." As we have said, they have a typical energy E_2 and are much weaker than those in charge of the linear memory. In the crudest theory, we may completely neglect them. Then we shall end up with exactly what is called an ideal polymer chain. This is just how we handled all the calculations in the previous chapters. It worked fairly well, and we coped with quite a number of problems. We described how a chain rolls up into a loose coil, and we revealed the peculiar entropic nature of the high elasticity of polymers.

Nevertheless, the ideal polymer chain approximation proves not to be enough for many other purposes. The properties of real polymers are much richer and more diverse than it predicts. If you are not convinced, think back to Chapter 3. There we talked about the various physical states of polymers. To understand

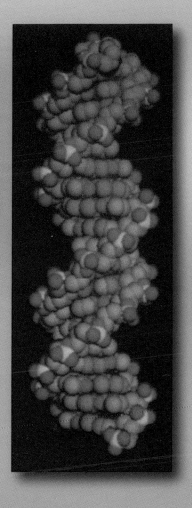
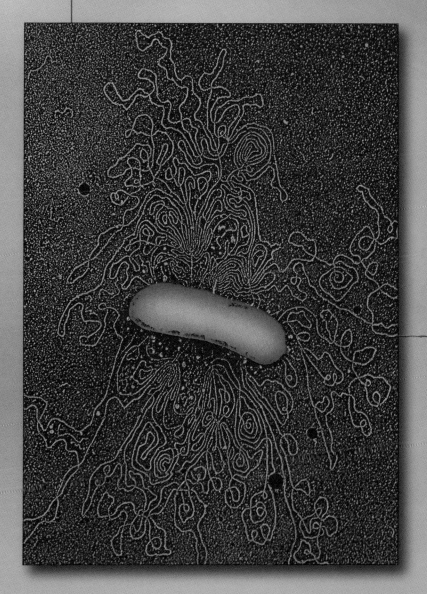

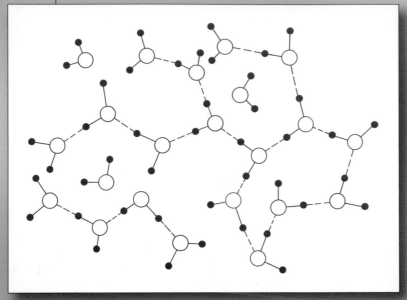

(a)

PLATE 3

The network of hydrogen bonds in water: (a) A simple back-of-the-envelope two-dimensional cartoon. (b) A more realistic computer-generated image (Section 4.1).

(*Source:* (b) D. Stauffer and H. E. Stanley, *From Newton to Mandelbrot*, New York: ©Springer-Verlag, 1990. Courtesy of H. E. Stanley. Reprinted with permission.)

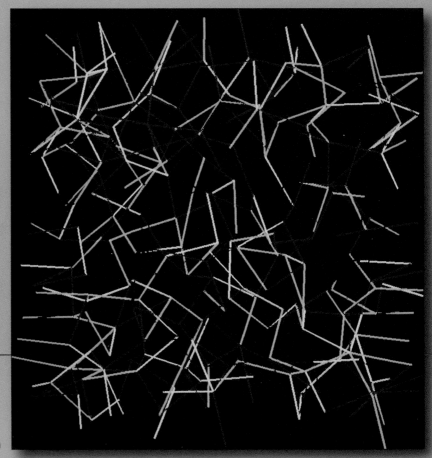

(b)

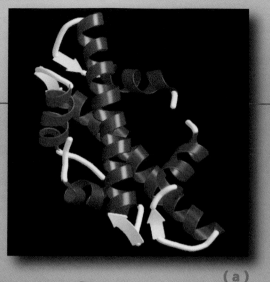

(a)

(b)

PLATE 4

Tertiary structures of two globular proteins: (a) Recombinant histone rHMfB. (b) The HIV-1 matrix protein (Section 4.6). (*Source:* (a) M. R. Starich, K. Sandman, J. N. Reeve, and M. F. Summers (1996), *J. Mol. Biol.*, v. 255, n. 1, 187–203; (b) M. A. Massiah, M. R. Starich, C. Paschall, M. F. Summers, A. M. Christensen, and W. I. Sundquist (1994), *J. Mol. Biol.*, v. 244, n. 2, 198–223.)

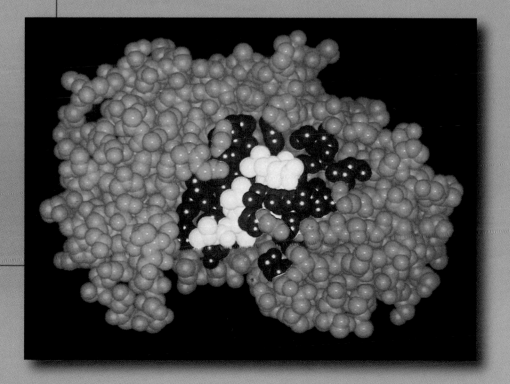

PLATE 5

A realistic model of molecular recognition. The main body of the protein globule is shown in blue. Yellow is the "target" molecule to be recognized. Red is the part of the protein that forms an active center (Section 4.7). (*Source:* J. Gomez and E. Freire (1995), *J. Mol. Biol.*, v. 252, n. 3, 337–350.)

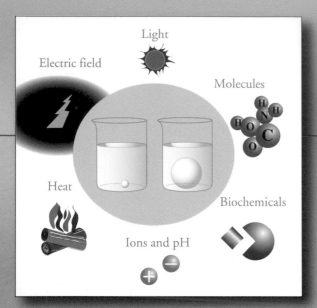

PLATE 6

A variety of factors that can cause gel to collapse (Section 8.11). (*Source:* T. Tanaka, original rendering.)

PLATE 8

Rings of relatively short DNA (Section 8.12). (*Source: Dictionary of Science and Technology,* Christopher Morris, ed., San Diego, CA: Academic Press, 1992.)

PLATE 7

Pattern on the surface of the gel sample in the course of swelling. There are innumerable jokes among the physicists doing polymer gels about these patterns being reminiscent of a brain (Section 8.11). To be serious, this pattern in fact has to do with the Norwegian coastline—see Plate 16. (*Source: Discover,* July, 1996, 10. Len Irish/© 1996, The Walt Disney Co. Reprinted with permission.)

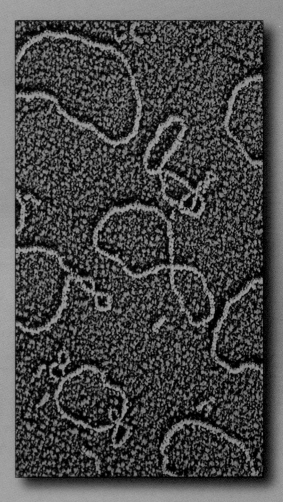

PLATE 9

While this illustration doesn't pretend to be very serious, it does explain why we compare the protein folding transition to "reading with understanding." Indeed, we are unable to read the sequence of letters when the polymer is stretched out, but when it collapses, it says clearly, "I am a protein" (Section 8.13).

PLATE 10

The minimum number of monomers for which a compact polymer can be knotted is 36. The conformation shown fills a 3x3x4 domain on the cube lattice, and has a knot. The monomers are shown as little cubes here and as spheres in Figure 8.13 and Plate 11, simply to stress that neither of these shapes is of any real significance and thus either can be used (Section 8.14)

PLATE 11

Lattice globule with an "active site" capable of specifically recognizing a "target molecule" (Section 8.14).

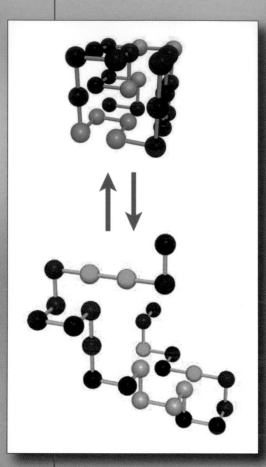

PLATE 12

If for some reason the globule shown in Plate 11 is denatured, it is able to renature, restoring the correct "active site." It is important that the globule does not need any assistance to renature, because in this sense it is indeed like a molecular machine (Section 8.14). (*Source:* V. S. Pande, original rendering.)

PLATE 13

A somewhat fictitious picture presenting many molecular machines working in the water medium (Section 8.14). (*Source:* V. S. Pande, original rendering.)

PLATE 14

Detail of a floor mosaic in a church in the village of Anagni, Italy, which was built in 1104. The mosaic is made of Serpinski gaskets (Section 10.2). (*Source:* D. Stauffer and H. E. Stanley, *From Newton to Mandelbrot*, New York: ©Springer-Verlag, 1990. Courtesy of H. E. Stanley. Reprinted with permission.)

PLATE 15

A cauliflower head and floret are similar to each other (Section 10.4).

(a)

(b)

PLATE 16

A map of Norway. The arrows point to the towns of Bodö and Tromsö (Section 10.4). (*Source:* Hammond world map, ©Hammond, Inc., Maplewood, NJ, license #12, 297.)

PLATE 17

DNA "walk" ($S(t)$ as a function of t) looks similar at different magnifications. Here, the DNA walk is shown for the piece of rat DNA called embryonic skeletal myosin heavy chain gene (Section 10.9). (a) represents the entire sequence; (b) and (c) are the magnifications of the solid boxes in (a) and (b) respectively. To obtain this obvious statistical self-similarity, one must magnify the segments by different factors along the t (horizontal) and the S (vertical) axes. These factors, M_t and M_S, respectively, are related to the scaling exponent α by the simple relation $\alpha = \ln(M_S)/\ln(M_t)$. For example, from (a) to (b) $\ln(M_S)/\ln(M_t) \approx \ln(2.07)/\ln(3.2) \approx 0.63$. (*Source:* H. E. Stanley *et al.* (1993), *Fractals*, v. **1**, n. 3, 49. ©World Scientific. Reprinted with permission.)

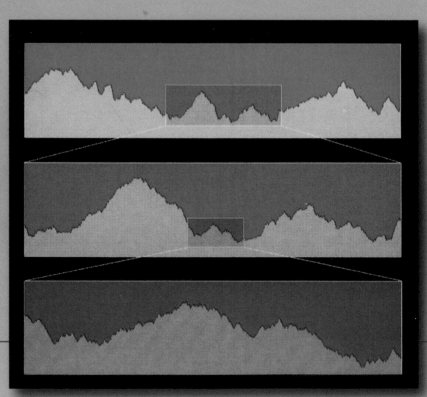

(a)

(b)

(c)

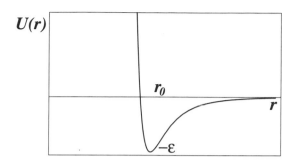

FIGURE 7.1
A typical dependence
of the interaction
energy *U* between two
particles on the
distance *r* between
them.

fully how all those states are formed, and why, we need to allow for excluded volume interactions. These include, in particular, interactions between different macromolecules. Monomers in the same macromolecule will also interact, even if they are not nearest neighbors but somehow come close to each other in space. In this chapter, we shall look at interactions of both types.

What can we say in general about an excluded volume interaction between two monomers? It certainly depends on the type of chain, as well as on the solvent. However, we can roughly sketch the potential energy of this interaction, $U(r)$, as a function of the distance r between the monomers (Figure 7.1).[1] It behaves in the usual way (given that nothing else is going on, like some additional strong interaction between the monomers). If r is small, $U(r)$ is positive and very large. This is because the monomers cannot penetrate into each other. In other words, the volume taken up by each monomer is automatically excluded from that available to any other one (hence the phrase "excluded volume"). As r becomes larger, monomers usually start to attract each other. This is the region on the right-hand side of the minimum in Figure 7.1. Clearly, the crossover distance r_0 between the two regimes (corresponding to the minimum) should have the same magnitude as the size of a monomer unit, that is, $r_0 \sim 10$ Å $\sim 10^{-9}$ m. What is the physical meaning of $U(r)$? To bring two monomers together, as close as r, some work has to be done. This work is stored in $U(r)$. It is done against the solvent molecules, as they need to be squeezed out of the way. Hence, the potential energy $U(r)$ represents the effective interaction of monomers through the solvent. It should depend, therefore, on the contents and state of the solvent, as well as on the temperature.

---

[1] The potential energy does not depend only on r, but also on the mutual orientation of the monomers. We do not take this into account directly, since the main qualitative features are well enough represented by the simplified Figure 7.1.

7.2 — Excluded Volume: Formulating the Problem

Let's discuss how interactions of the type shown in Figure 7.1 might influence the shape of an isolated polymer chain in a dilute solution (Figure 3.7*a*). First of all, would excluded volume interactions make the coil swell or shrink? This, it turns out, depends on the temperature of the solution.

Suppose the characteristic energy of attraction ε (Figure 7.1) is much greater than the thermal energy kT. Then attraction will dominate. As a result, the macromolecule will shrink to become more compact than an ideal coil. This is a special polymer state, called a polymer globule. We shall come back to it in the next chapter.

It is not the same story if ε is smaller than kT. In this case, attraction is not too important. Repulsion at shorter distances between monomers is the prevailing form of interaction. It makes the coil swell. Such swelling is called the excluded volume effect. (You presumably understand where the name comes from. As we have already said, the repulsion at short distances occurs because the volume of each monomer is excluded for all the others.) In this chapter, we are going to tackle the problem of excluded volume; that is, we shall try to picture how a polymer coil swells.

For an isolated polymer chain, the problem is purely geometrical. Indeed, the spatial shape of an ideal chain resembles the path of a randomly wandering Brownian particle (see Chapter 5). What new features will the shape of the chain acquire if we allow for the excluded volume? Clearly, since the "private space" of each monomer is not available to the rest, the chain cannot possibly cross itself at any stage. This sort of behavior can be described as self-avoiding. For example, if there were an equivalent Brownian particle, it would not be allowed to cross its own track. A two-dimensional version of such a trajectory is sketched in Figure 7.2. Thus, we have made it a purely geometrical problem of self-avoiding random walks.

This problem can be quite successfully approached by computer simulation. The simplest way to set it up is to use a random number generator to try out various trajectories of a polymer chain (just as described in Section 2.4). Then whenever we obtain with a self-crossing trajectory, we merely ignore it. Thus, we only keep the self-avoiding paths, and when we have enough of them, we can look at some average features. Although more sophisticated algorithms are normally used these days, in principle they are not that different from what we have just described. A typical result is shown in Figure 7.3.

So what has been gathered from the computer simulations of self-avoiding walks? It appears that the conformational properties of a polymer coil are quite

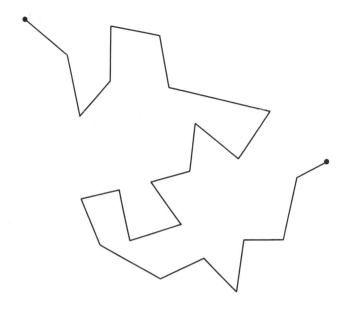

FIGURE 7.2
A self-avoiding path in two dimensions (on the plane).

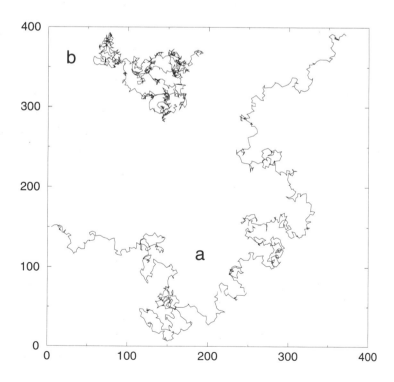

FIGURE 7.3
(*a*) A typical Gaussian conformation of a polymer of 1000 monomers that is self-avoiding in three-dimensional space. Each segment length is 1. Although in three dimensions a self-avoiding chain never crosses itself, on the plane two-dimensional projection crossings are possible. This is why this figure is very different in spirit and in its meaning from Figure 7.2. (b) A typical Gaussian conformation for the same chain is shown for comparison. *Source:* Courtesy of S. Buldyrev.

significantly affected by the excluded volume. The coils become looser, and the fluctuations in the segment concentration become more severe. The mean-square size of the coil increases. Moreover, the mean-square end-to-end distance $\langle R^2 \rangle$ now depends differently on the number of segments in the chain. Instead of the familiar $\langle R^2 \rangle \sim N$ (which we derived for an ideal chain in Chapter 5), we now get the following results. For self-avoiding walks in three dimensions,

$$\langle R^2 \rangle \sim N^{6/5}. \tag{7.1}$$

For self-avoiding walks on a plane (i.e., in two dimensions),

$$\langle R^2 \rangle \sim N^{3/2}. \tag{7.2}$$

Of course, the exact fractions $\frac{6}{5}$ and $\frac{3}{2}$ do not appear as such in the computer simulations. But within a certain accuracy, the indices produced by a computer are close to these values. Relationships (7.1) and (7.2) confirm that a polymer coil with excluded volume is swollen compared to an ideal coil. You may find it handy to introduce a swelling coefficient α, such that

$$\alpha^2 = \frac{\langle R^2 \rangle}{\langle R^2 \rangle_0} \tag{7.3}$$

(where $\langle R^2 \rangle_0 = N\ell^2$ is the size of an ideal polymer chain). As N increases, α grows as $N^{1/5}$ in three dimensions and as $N^{1/2}$ in two dimensions.

Polymer swelling is illustrated in the special movie called "Swelling" on the CD ROM. We begin with a Gaussian polymer coil and gradually increase the excluded volume of the monomers. This corresponds physically to a gradual change in the solvent quality. One can see that, in a good solvent, a polymer does indeed swell compared with its Gaussian size.

7.3 — The Density of a Coil and Collisions of Monomer Units

The problem of excluded volume did not succumb to the efforts of theorists for more than 20 years. The way to tackle it, or, to be more precise, the way to reduce it to some other, better explored problems, was found by P. G. de Gennes in 1972. His solution goes far beyond what we can explain in this book. However, we do not have to jump straightaway to equations (7.1) and (7.2). To convince you, we offer a simple explanation known as Flory's theory, although the way we are going to present it does not quite follow Flory's original version.

First of all, the mean spatial size of a coil R is obviously of the same order of magnitude as $\langle R^2 \rangle^{1/2}$. This is why we can write that $R \sim \ell N^{1/2}$ for an ideal polymer (see (5.11)). Meanwhile, for a coil with excluded volume we get $R \sim \alpha \ell N^{1/2} > \ell N^{1/2}$. The volume taken up by such a coil is $V \geq \ell^3 N^{3/2}$ (we omit the factor $4\pi/3$ as usual). However, a polymer chain never uses up all the space inside the coil. This is clearly seen in Figure 2.6. You can show it is true by the following argument. Let the volume of a single monomer segment be v. Then the total volume of the coil is Nv. Since $N \gg 1$, we have $V > \ell^3 N^{3/2} \gg Nv$. In other words, the fraction ϕ of the volume of the coil taken up by the monomer segments is really very small:

$$\phi \sim \frac{Nv}{V} < \frac{Nv}{\ell^3 N^{3/2}} \sim N^{-1/2} \left(\frac{v}{\ell^3} \right) \ll 1. \tag{7.4}$$

(In Section 5.6, we used the same sort of argument for an ideal polymer.) The same thing can be said about the mean concentration of the segments in the coil, $n \sim N/V < \ell^{-3} N^{-1/2}$ (cf. (5.14)). At first glance, you may think it implies that a polymer with excluded volume is always ideal. Indeed, if the segment concentration is so low, their encounters are very rare, and one can be tempted to neglect them. On the other hand, we know that the coil is very pliable, and its elastic modulus is small.

This suggests that the question should be treated with more subtlety. Let's make a crude estimate of how many encounters (i.e., collisions) between the segments of the coil may occur at the same time. Assume that the coil is a cloud of totally independent particles (segments) spread over the volume V. It would be wonderful if we could take a three-dimensional photo of this cloud. We would then be able to count all the collisions between two, three, or more bodies, caught at a moment in time.

Unfortunately, we cannot do this, so we have to use another approach. There are N particles all together. The probability that each particle has a close "partner" is ϕ. This is why the number of pair collisions is of order $N\phi$. In the same way, the number of three-body collisions is roughly $N\phi^2$, and so on. In general, the number Y_p of p-body collisions can be estimated as $N\phi^{p-1}$. From (7.4),

$$Y_p \sim N\phi^{p-1} < N^{(3-p)/2} \left(\frac{v}{a^3} \right)^{p-1}. \tag{7.5}$$

You can see that $Y_p \simeq 1$ if $p > 3$. This indicates that many-body collisions are really rare. Even the number of three-body collisions in a swollen coil is of order 1. So they cannot seriously affect the conformation of the coil. In contrast, the number of simultaneous pair collisions is about $N^{1/2}$. This is much less than N (so each particular segment seldom has a collision), yet it is a large number compared to 1.

Besides, as we showed in Chapter 6, a long polymer chain is very pliable, and its elastic modulus is small ($\sim 1/N$, see (6.21)). Therefore, we have the right to suspect the pair collisions of making the polymer swell in the way implied by (7.1) and (7.2).

What is the free energy of a polymer like given the excluded volume interactions? (See (6.19).) Of course, it has the usual entropy term $-TS$ (which would be the only term in the case of an ideal gas). In addition, it includes the internal energy U of the segment interactions. This latter term is responsible for the swelling. In other words, it accounts for the excluded volume effect. All we need to know now is the contribution of the binary collisions to the internal energy U of the coil.

Here is how we can find it. The segment density n is very low, as we have seen. So U can be expanded as a series of powers of n:

$$U = VkT \left[n^2 B + n^3 C + \cdots \right], \tag{7.6}$$

where V is the volume of the coil and B and C are expansion coefficients, or virial coefficients (i.e., B is the second virial coefficient, C is the third, and so on). These coefficients are fully determined by the form of the interaction potential $U(r)$ and the temperature T. Obviously, the first term in expansion (7.6) stands for the binary interactions. This is because it is proportional to n^2, which is just the pair collision probability. Likewise, the second term is related to three-body interactions, and so on.[2] Thus, the energy of all the binary interactions between the segments of a coil is

$$U = VkTn^2 B, \tag{7.7}$$

where n is the average segment density in the coil.

---

[2] Here is a useful exercise for your leisure time. Using (7.6), derive the equation of state for an ordinary imperfect gas (i.e., the relationship between volume, pressure, and temperature). Say the volume is V and the number of molecules is N. Then $n = N/V$. The internal energy U is given by (7.6), and the entropy S and the free energy F can be found from (5.10) and (5.15). You can work out the pressure by differentiation: $p = -(\partial F/\partial V)$. Check that this leads to the Clapeyron–Mendeleev equation $pV = NT$ given that $B = 0$ and $C = 0$. For nonzero B and C, compare your answer with the van der Waals equation $((p + a/V^2)(V - b) = NT)$. Prove that they are qualitatively the same for gases of moderate density. Good luck!

7.4 ─ Good and Bad Solvents, and Θ Conditions

We have already discussed the potential $U(r)$ in Figure 7.1. Repulsion between the segments dominates at higher temperatures ($\varepsilon \ll kT$) (the excluded volume effect), whereas at lower temperatures ($\varepsilon \gg kT$) attraction takes over. Let's look at the higher temperature region first. The most important values of r are those where $U(r) > 0$. So the internal energy of the coil (as well as the second virial coefficient) is positive. In contrast, at lower temperatures it is the "attractive" part of $U(r)$ that gives the biggest contribution. This is where $U < 0$. So both U and B are negative. In the former case we say that we are dealing with a good solvent and in the latter case with a bad one. We are not being biased! If in a solvent the segments of polymer chains tend to repel each other, the polymer will dissolve. Conversely, if the segments attract each other, the polymer chains will be rather "sticky"; in other words they will stick together and precipitate out rather than dissolve.

The quality of a solvent (i.e., whether it is good or bad) may change with its contents or with temperature. Hence, there has to be a special point where the second virial coefficient goes through zero: $B = 0$. It is usually called the Θ point (or Θ temperature—obviously this is the temperature when $B = 0$). At the Θ point, attraction and repulsion between the segments completely cancel out, and the behavior of the polymer becomes ideal. When $T > Θ$, repulsion dominates. This is the excluded volume (and good solvent) region. In contrast, when $T < Θ$ attraction prevails, making the solvent bad. We can now rephrase our initial problem. The swelling of a polymer due to the excluded volume effect is the same as the swelling of a polymer in a good solvent, that is, at $T > Θ$.

You may wonder why such conditions are possible in the first place. Is it a mere coincidence that at a certain point repulsion and attraction are so perfectly balanced? The answer is that the cancelation only works because three-body interactions (and all the higher ones) are not important. Their contribution to U is always very small. As for the binary collision term (7.7), it is proportional to B, so it falls to zero at the Θ point. Hence, all that really remains of U at $T = Θ$ is the entropy term (see (6.19)). This is why the coil's behavior becomes ideal.

Thus, the existence of the Θ point (where the segment interactions have no influence on the shape of the chain) is yet another peculiarity of polymers. It is all to do with the very low segment density $n \sim N^{-1/2}$. An imperfect gas, for instance, is far from becoming ideal when $B = 0$. Many-body collisions are not negligible, and there is nothing to counteract them. In the case of a gas, there is no temperature at which attraction and repulsion cancel out.

7.5 — The Swelling of a Polymer Coil in a Good Solvent

Let us consider an isolated polymer coil in a good solvent ($B > 0$) and try to find its swelling coefficient α. The first calculation of this sort was done by P. Flory in 1949. His approach was as follows. The main cause of the swelling is repulsion between the segments inside the coil (the binary collisions). However, there is also an effect that hinders swelling, arising from the elastic forces whose origin is due to entropy (we discussed them in Chapter 6). These forces emerge because there are fewer different shapes that the chain can take when it is straightened out (or swollen). So Flory's idea was to obtain the swelling coefficient α from a balance condition between the repulsive and elastic forces.

Both factors contribute to the free energy of a swollen polymer coil (with swelling coefficient α), $F(\alpha) = U(\alpha) - TS(\alpha)$ (see equation (6.19)). The potential energy term $U(\alpha)$ is determined by the repulsive interactions (see (7.7)):

$$U(\alpha) = VkTn^2B \sim \frac{kTR^3BN^2}{R^6} \sim \frac{kTBN^{1/2}}{\ell^3\alpha^3}. \tag{7.8}$$

To write (7.8), we used the following straightforward relationships: $V \sim R^3$, $n \sim N/R^3$, and $\alpha = R/R_0 = R/N^{1/2}\ell$, where $R_0 \sim N^{1/2}\ell$ is the size of an ideal polymer coil. As usual, we leave out numerical factors since we are only making estimates. Likewise, the entropy term $S(\alpha)$ in the free energy of the swollen coil results from elastic forces. We can work it out from (6.4):

$$S(\alpha) = \text{const} - k\frac{3R^2}{2N\ell^2} = \text{const} - k\frac{3N\ell^2}{2N\ell^2}\alpha^2 = \text{const} - \frac{3}{2}k\alpha^2. \tag{7.9}$$

The free energy $F(\alpha)$ is obtained from (7.8) and (7.9):

$$F(\alpha) = U(\alpha) - TS(\alpha) = \text{const} + K\frac{kTBN^{1/2}}{\ell^3\alpha^3} + \frac{3}{2}kT\alpha^2, \tag{7.10}$$

where const is independent of α and K is a constant of order one (it is just the numerical factor in (7.8) that we left out).

The function $F(\alpha)$ is sketched in Figure 7.4. You can see a minimum in the curve at a certain α. The minimum of the free energy always gives the equilibrium state. So the equilibrium swelling coefficient is just the value of α at the minimum.

Can we find out where exactly the minimum is? The usual way is to differentiate $F(\alpha)$ with respect to α and to set the derivative equal to zero. We

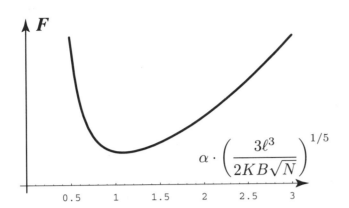

FIGURE 7.4
The dependence $F(\alpha)$ given by equation (7.10).

obtain

$$\frac{\partial F}{\partial \alpha} = -3K \frac{kTN^{1/2}B}{\ell^3 \alpha^4} + 3kT\alpha = 0. \qquad (7.11)$$

From here,

$$\alpha^5 = \frac{KBN^{1/2}}{\ell^3} \quad \text{i.e.,} \quad \alpha \sim \left(\frac{N^{1/2}B}{\ell^3}\right)^{1/5}. \qquad (7.12)$$

In the latter formula we used the sign $\sim$ (i.e., "of order") because we dropped the numerical factor K. We would not be able to find it anyway within this simple theory. All that Flory's theory really gives are the power indices in equations (7.1) and (7.2). From (7.12), the size R of the coil is estimated as

$$R \sim \alpha R_0 \sim \alpha N^{1/2}\ell \sim \ell N^{3/5} \left(\frac{B}{\ell^3}\right)^{1/5}. \qquad (7.13)$$

This agrees completely with the results (7.1) of computer simulations.

Using a similar method, we can also explain equation (7.2), which is the two-dimensional equivalent. Fortunately, the expression (7.9) for $S(\alpha)$ remains the same. As for $U(\alpha)$, we need to be careful. The two-dimensional "volume" is not $V \sim R^3$ as usual, but $V \sim R^2 \sim \alpha^2 N\ell^2$. Therefore, instead of (7.8) we arrive at

$$U(\alpha) \sim \frac{kTR^2BN^2}{R^4} \sim \frac{kTBN}{\ell^2 \alpha^2}. \qquad (7.14)$$

The free energy is calculated as $F(\alpha) = U(\alpha) - TS(\alpha)$, where $U(\alpha)$ and $S(\alpha)$ are given by formulas (7.14) and (7.9), respectively. Now we can work out where $F(\alpha)$ reaches a minimum, using the same idea as before. The answer is

$$\alpha \sim \frac{BN}{\ell^2}. \tag{7.15}$$

Finally,

$$R \sim \alpha R_0 \sim \alpha N^{1/2}\ell \sim \ell N^{3/4}\left(\frac{B}{\ell^2}\right)^{1/4}, \tag{7.16}$$

in total agreement with (7.2).

Thus, if we allow for the excluded volume effect, the average size of a polymer coil will no longer be proportional to $N^{1/2}$ (as for an ideal chain), but to $N^{3/5}$ in three dimensions and to $N^{3/4}$ in a plane. So, just as we expected, the excluded volume effect is quite significant. This is despite the coil being extremely loose and the collisions between the segments being very unlikely. Indeed, if $N \to \infty$ the swelling coefficient also grows without limit. The analogy with a Brownian particle no longer makes sense for a swollen coil. If a Brownian particle were not allowed to cross its own path, it would move much farther away from its starting point in a given time.

7.6 — The Excluded Volume Effect in a Semi-dilute Solution

As we discussed in Section 3.6, isolated polymer coils are typical for dilute solutions, where the volumes taken up by the coils do not overlap (Figure 3.7a). Things change when the polymer concentration exceeds the threshold value $c^\star$ (which is defined by equation (5.14) for an ideal polymer). In this case we have a semidilute solution (Figure 3.7c). Although the fraction of the volume taken up by the polymer is still rather small, the coils are already highly intermingled. Can we work out what the excluded volume effect does to the coils in this case (i.e., when the polymer concentration $c \gg c^\star$)?

First of all, we need to know the value of $c^\star$ for a semidilute solution. When $c = c^\star$, the average segment density in the whole solution becomes equal to the average segment concentration inside each coil (see Section 5.6). Therefore, taking into account the excluded volume, we get

$$c^\star \sim \frac{N}{R^3} \sim \ell^{-3} \left(\frac{B}{\ell^3}\right)^{-3/5} N^{-4/5}. \tag{7.17}$$

Since $N \gg 1$, the threshold density $c^\star$ is fairly small (just as it was for an ideal polymer solution; see (5.14)). Thus the semidilute regime is appropriate for a wide range of concentrations.

In the same way as for a dilute solution, we can describe the swelling of the coils by means of $\langle R^2 \rangle$. (As always, $\vec{R}$ is the end-to-end vector of the coil.) Obviously, the average size of the coil R is estimated as $R \sim \langle R^2 \rangle^{1/2}$. Now we need to calculate the value $\langle R^2 \rangle$ in a semidilute solution.

Choose a certain polymer chain and fix in space one of its monomer units. Now look at a strand of the chain near the fixed unit. Suppose this strand contains g monomers. If there were no other chains around, the excluded volume effect would swell the strand roughly to a size $\ell g^{3/5}(B/\ell^3)^{1/5}$ (see (7.13)). The volume taken up by such a g-strand would be of order $\left[\ell g^{3/5}(B/\ell^3)^{1/5}\right]^3 \sim \ell^3 g^{9/5}(B/\ell^3)^{3/5}$. The monomer concentration in this volume would be estimated as $g/\left[\ell^3 g^{9/5}(B/\ell^3)^{3/5}\right] \sim \ell^{-3}g^{-4/5}(B/\ell^3)^{-3/5}$. It decreases as g is increased. This is not surprising. Since the neighbors are tied to the fixed monomer with a piece of polymer chain, there is a sort of "correlational" thickening around this area. (We call it "correlational" because it results from interactions or correlations between the monomers in the chain.) We must not forget that this strand is not alone. It is in an ocean of intermingled, overlapping polymer chains, with monomer concentration c. Do the surrounding chains manage to penetrate into the densest region near the fixed monomer? The answer is no. There is just no room, since the monomers cannot go through each other (the excluded volume effect). In this region, the correlational density is higher than c. We can find the size $\xi^\star$ of this region (i.e., where there is correlational thickening) and the number of monomers in it $g^\star$ from the following conditions: $\ell^{-3}(g^\star)^{-4/5}(B/\ell^3)^{-3/5} \sim c$, and $\ell (g^\star)^{3/5}(B/\ell^3)^{1/5} \sim \xi^\star$. Hence,

$$\xi^\star \sim \ell(c\ell^3)^{-3/4}\left(\frac{B}{\ell^3}\right)^{1/4} ; \quad g^\star \sim \left(c\ell^3\right)^{-5/4}\left(\frac{B}{\ell^3}\right)^{-3/4}. \tag{7.18}$$

Note that (7.17) leads to $g^\star < N$ provided that $c > c^\star$.

Now that we have a clearer picture, let's draw some conclusions. In the case of a semidilute solution ($c > c^\star$), it helps if we divide each chain into a sequence of strands, or blobs, of a certain length $g^\star$. Each blob, taken separately, looks like a normal isolated polymer chain, swollen by the excluded volume effect. The size $\xi^\star$ of the blob provides an important length scale. That is, there is no correlation between the monomers of the chain at distances longer than $\xi^\star$. You could say that, due to the excluded volume effect, the "inner lives" of neighboring blobs are screened from each other. So a blob in a semidilute solution does not really "care" whether the neighboring blobs belong to the same chain as itself, or not. Now, this is interesting. Let's zoom out and look at a chain at a less detailed scale. We shall see a sequence of blobs. What if we regard the blobs as new, bigger monomer units? The chain of blobs will behave as an ideal one, so we can apply the usual theory of a Gaussian chain. The number of blobs in the chain is about $N/g^\star$, and the size of each of them is of order $\xi^\star$. Therefore, we have

$$R \sim \langle R^2 \rangle^{1/2} \sim \xi^\star \left(\frac{N}{g^\star} \right)^{1/2} \sim \ell N^{1/2} (c\ell^3)^{-1/8} \left(\frac{B}{\ell^3} \right)^{1/8}. \qquad (7.19)$$

Figure 7.5 gives an idea of how R depends on the concentration c of the solution. When $c < c^\star$, the size of the coil is not affected by c; it is simply described by equation (7.13). If c is increased, we shall reach the regime $c > c^\star$, that is, a semidilute solution. Here the swelling coefficient starts dropping, just as (7.19) predicts. By the way, for $c \sim c^{\star\star} \sim B/\ell^6$, equation (7.19) leads to $R \sim N^{1/2}\ell$. Thus, when the density becomes very high, the excluded volume effect no longer causes swelling. This conclusion refers, in particular, to polymer melts, where there is no solvent at all and c reaches the highest possible value. If you still remember the Flory theorem (Section 6.8), you will not be surprised by this.

FIGURE 7.5
A sketch of the dependence $R(c)$.

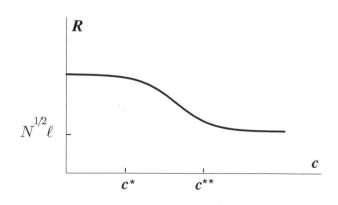

7.7 — The Compatibility of Polymer Blends

We will finish this chapter with a section on the compatibility of polymer blends. It is not directly related to the problem of the excluded volume, but we have a "debt" left over from Section 3.7. We have not yet proved that, if contact between monomers of types $\mathcal{A}$ and $\mathcal{B}$ in a polymer $\mathcal{A}$/polymer $\mathcal{B}$ blend is even slightly energetically unfavorable, phase separation into almost pure $\mathcal{A}$ and $\mathcal{B}$ phases will occur. Now, at last, we are equipped with all we need to prove this.

Let us take a mixture of $\mathcal{A}$ and $\mathcal{B}$ monomers that are not linked into chains. Can we compare the phase separation in this mixture (Figure 7.6a) with that in a polymer blend (Figure 7.6b)? In each, the number of energetically unfavorable $\mathcal{A}$–$\mathcal{B}$ contacts drops dramatically during the phase separation into almost pure $\mathcal{A}$ and $\mathcal{B}$ phases. These contacts may only take place along the surface separating the two phases. We can say that the gain in energy due to the phase separation is the same in both cases.

However, the phase separation not only gains energy, but also leads to a loss of entropy. This is because the number of possible conformations of the system is lessened. When mixed, $\mathcal{A}$ and $\mathcal{B}$ molecules had access to the whole volume. Once separated, they can only reach a part of it (cf. Section 6.7). Are these losses in entropy the same for Figures 7.6a and 7.6b? The answer is no due to how many possible conformations can be realized in either case. Obviously, this number is many orders of magnitude greater in a low molecular weight mixture of $\mathcal{A}$ and $\mathcal{B}$ than in a polymer blend. Unattached monomers in the mixture can move

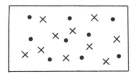
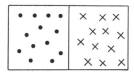

a

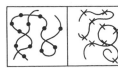

b

FIGURE 7.6
Phase separation in (a) a low molecular weight mixture and (b) a polymer blend. $\mathcal{A}$ **and** $\mathcal{B}$ **components are shown with circles and crosses respectively.**

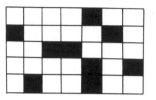

a

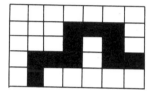

b

independently of each other, whereas in the case of the polymer blend they are linked into chains.

To give an illustration, we will calculate the number of possible conformations of N molecules on a square lattice containing V cells (Figure 7.7). We shall consider two cases: N independent molecules, each taking up just one cell (Figure 7.7*a*), and N molecules linked together into a chain (Figure 7.7*b*).

In the first case, the answer is obvious. There are V possibilities for each molecule. (Let us assume that $N \simeq V$ and neglect the excluded volume of the molecules.) Therefore, the total number of conformations is V^N.

Now, let's look at the second case. The first monomer unit can be positioned in V different ways, the second one in four ways (i.e., in the cells sharing edges with the first), the third and all the following ones in three ways (Figure 7.7*b*). Thus, we get $4V \cdot 3^{N-2}$ ways altogether.

Obviously, if $N \gg 1$,

$$V^N \gg 4V \cdot 3^{N-2}. \tag{7.20}$$

This proves that the number of conformations is much less, and therefore the entropy is much lower for the polymer system.

This is why the entropy losses due to the phase separation are much smaller in a polymer blend (Figure 7.6*b*) than in the corresponding low molecular weight system (Figure 7.6*a*). However, the energy gain is the same (see above). Thus, it is the energy that dominates. Even a very weak repulsion between $\mathcal{A}$ and $\mathcal{B}$ monomers is enough to compensate for some minor entropy losses due to the phase separation. To see how weak the repulsion can be, suppose you mix two polymers, $\mathcal{A}$ and $\mathcal{B}$, containing N units each. Theory shows that the contact energy ε between them, which is sufficient to induce the phase separation, can be estimated as

$$\varepsilon \sim \frac{kT}{N}. \qquad (7.21)$$

Clearly, for large N, this value is really very low.

8

Coils and Globules

He may be only little,
But he's a good boy.
V. Mayakovsky, *What Is Good And What Is Bad?* (Russian children's poem)

What are little boys made of, made of?
...
What are little girls made of, made of?
...
The Mother Goose rhymes.

8.1 — What Is a Coil-Globule Transition?

In the previous chapter, we focused on the excluded volume problem. We learned to appreciate that each monomer has a certain volume, and the monomers cannot penetrate each other. This leads to repulsion at short distances. In the case of a good solvent, repulsion is the prevailing tendency overall, so the polymer coils swell. But what if the quality of the solvent grows worse? For example, you could add some precipitant into the solvent or change the temperature. As a result, the solvent may go through the Θ point (as we discussed in Section 7.4) and the binary interactions between the monomers will become mainly attractive. Segments will tend to stick to each other from time to time, so there will be lots of temporary couples. What will this do to the coil as a whole?

W. Stockmayer was the first to predict, in 1959, that if the attraction between monomers becomes strong enough, the polymer undergoes a phase transition of the same sort as the transition from gas to liquid. Bits of the polymer "condense on to themselves" and instead of a loose coil you end up with a dense "drop"—a polymer globule. This is just what is meant by coil-globule transition.

A typical globule with strongly interacting monomers is portrayed in Figure 8.1. This picture was obtained by computer simulation. (Compare it with Figure 2.6, which shows an ideal polymer coil.) How is attraction modeled on a

127

FIGURE 8.1
(*a*) A typical
conformation of a
polymer globule (in the
top left corner) and
(*b*) a polymer coil (with
excluded volume). Both
have been generated
computationally for the
polymer of 1000
segments, of the length
1 each. *Source*: Courtesy
of S. Buldyrev.

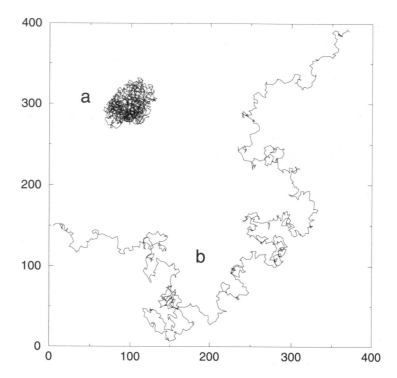

FIGURE 8.1
(*a*) A typical
conformation of a
polymer globule (in the
top left corner) and
(*b*) a polymer coil (with
excluded volume). Both
have been generated
computationally for the
polymer of 1000
segments, of the length
1 each. *Source*: Courtesy
of S. Buldyrev.

computer? First, you can figure out the attractive potential energy of the binary interactions (of the same kind as shown in Figure 7.1). Then you will know the forces of interaction. Then you can use Newton's laws and trace, on a computer, how the chain would move under these forces. You will see some shapes similar to Figure 8.1. In contrast to the coil, the globule is very dense and compact, and there are no vast "holes" inside it. The only real difference between a globule and an ordinary liquid is that the globule's "molecules" (i.e., the monomers) are all linked together.

Polymer globules and coil-globule transitions came in from the cold thanks to molecular biophysics. Some of the most important biological polymers—protein enzymes—usually appear in living cells in globular form (we mentioned them in Chapter 4). If something nasty happens to the solvent surrounding the proteins (say, it gets overheated, or the contents are disturbed), the proteins may be denatured. In other words, they lose all their biochemical activity. Denaturation of proteins usually implies a dramatic change in shape and is accompanied by a strong absorption of heat. The first scientists who worked on the coil-globule transition were inspired by the thought that it might shed some light on the denaturation of proteins. It seemed quite plausible that when denaturation occured, the dense globular structure would be destroyed and the protein would take on the shape of a coil.

Only later did it turn out that there is no straightforward analogy between the coil-globule transition and protein denaturation. However, the coil-globule transition was quite extraordinary and exciting in its own right. This stimulated further interest in the problem. The studies have expanded, covering all kinds of polymers. A globular state has been discovered for many other systems. Not only proteins, but also DNA molecules and macroscopic polymer networks, for example, can have a globular structure. This explains quite a few unusual polymer effects, such as the existence of the compact form of DNA and the so-called collapse of polymer networks. We shall talk about some of them a little later. This very broad view of the coil-globule transition was initiated in 1968 by the pioneering work of the Russian physicist I. M. Lifshitz.

8.2 — The Free Energy of a Globule

Let's now take an isolated polymer molecule and try to build a simple theory for the coil-globule transition. For our purposes, we do not need to go into the details of Lifshitz's consistent approach. Instead, we will stick to the same logic as when we used Flory's approximate arguments to tackle the excluded volume problem (see Section 7.5). As you remember, we introduced the swelling coefficient α; the free energy of a swollen coil was written as the sum of two terms (see (7.10)). One term was the free energy associated with stretching the coil by the factor α, $U_{eff} = -TS(\alpha)$. The other term was the energy of the monomer interactions in the coil, $U(\alpha)$. Then we found the equilibrium value of α; it was the α corresponding to the minimum of the free energy $F(\alpha)$.

Where did the term $U_{eff}(\alpha)$ come from? It has to do with the poorer choice of shapes that a straightened polymer can take. Fewer possibilities means lower entropy and a lower probability of the elongated state (see Section 6.5). Meanwhile, we have already said that the other term, $U(\alpha)$, is the energy of the monomer interactions. Thus, when we write the free energy in the form (7.10), we automatically distinguish its entropy and energy parts. Notice that we never use the fact that $\alpha > 1$. It applies just as well if the molecule shrinks ($\alpha < 1$) instead of swells ($\alpha > 1$). We still have the free energy in the same form:

$$F(\alpha) = U_{eff}(\alpha) + U(\alpha). \tag{8.1}$$

Here $U_{eff}(\alpha)$ is determined by the entropy of the final state of the coil, when it is either swollen or shrunk by the factor α. (Although in the case of shrinking α is less than one, we would like to keep its previous name, the "swelling coefficient." After all, its mathematical definition (7.3) is still the same.)

8.3 — The Energy of Monomer Interactions

If we want to cater for both regimes—a swollen coil ($\alpha > 1$) and a shrunken globule ($\alpha < 1$)—we need to express the two terms in (8.1) in a slightly different fashion than Section 7.5. Consider the energy $U(\alpha)$ first. We used to estimate it, as in (7.8), by taking into account only the binary interactions between monomers (described by the second virial coefficient B). We had the right to do this, since the monomer density is very low in both ideal and swollen polymers. However, when a polymer shrinks (i.e., $\alpha < 1$), the monomer density goes up. Many-body collisions may now become important. This is why we can no longer stop after just the first term in expansion (7.6). Let's see what we gain if we keep the second term too. We shall have

$$U(\alpha) \sim R^3 kT \left(Bn^2 + Cn^3 \right) \tag{8.2}$$

$$\sim R^3 kT \left[B \left(\frac{N}{R^3} \right)^2 + C \left(\frac{N}{R^3} \right)^3 \right]$$

$$\sim kT \left[\frac{BN^{1/2}}{(\alpha^3 \ell^3)} + \frac{C}{(\alpha^6 \ell^6)} \right]$$

(cf. (7.8)), where $R \sim \alpha N^{1/2} \ell$ is the size of the molecule and C is the third virial coefficient, which represents three-body interactions. (In estimate (8.2) we have dropped all numerical coefficients of order of 1.) Will it do just to include the three-body interactions and ignore the rest? The answer is yes. If we did a more detailed calculation, we would see that the first two terms in expansion (7.6) are enough to give a correct account of the coil-globule transition. In principle, higher terms would be needed to describe the actual state of a dense globule. However, as we shall show later, a globule tends to swell before turning into a coil. So during the actual transition a globule's density is not so high.

What are the signs of the two terms in (8.2) in the transition region? Since the transition can only happen in a bad solvent (i.e., below the Θ temperature), the second virial coefficient $B < 0$, and the binary interactions are mainly attractive. As for the third virial coefficient C, it turns out normally that $C > 0$ in the transition region. So repulsion is the predominant type of three-body collision. In general, the higher the order of interaction, the wider the range where they are effectively repulsive. Roughly, we can explain this in the following way. Suppose a particle (or a monomer) interacts with a clump of m other particles. The excluded volume for this particle will be proportional to m. However, attraction emerges only in the surface layer. The volume of this layer is proportional to $m^{2/3}$. This is why, as long as m is large enough, repulsion will always prevail. It is actually quite fortunate—otherwise no substance would be stable, and all things would shrink without limit! To conclude, the energy $U(\alpha)$ can indeed be approximated by (8.2); in particular, it works in the case $B < 0, C > 0$, which we are considering in this chapter.

8.4 — The Entropy Contribution

Now let's concentrate on the entropy contribution $U_{eff}(\alpha)$ to the free energy. In the case of a swollen coil, this contribution was described by equation (7.9). Would it be valid for $\alpha < 1$ as well? Let's think. Equation (7.9) gives the free energy of an ideal coil whose end-to-end distance is of order $R \sim \alpha N^{1/2}\ell$. This is the only condition on the coil's shape that we used when deriving (7.9). Now suppose we have a shrinking coil ($\alpha < 1$). In this case, not only does the end-to-end distance stay roughly the same, but the whole coil needs to be fitted into a volume of linear size R (see Figures 8.2a and 8.2b). Hence, equation (7.9) would be no good in the case $\alpha < 1$. It would seriously underestimate the entropy loss.

So how can we find a reasonable estimate for $U_{eff}(\alpha) = -TS(\alpha)$? Let us look at the Boltzmann equation (6.2). It suggests that the entropy (as well as the

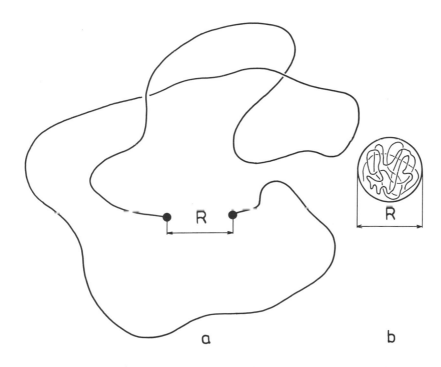

FIGURE 8.2
(a) An ideal polymer chain with a short end-to-end distance R, (b) a polymer chain squashed to the size R in all directions.

R

R

a

b

entropy loss) should not depend on the actual cause of the shrinking. This is good news. We no longer have to worry about a real polymer with some attractive forces between the monomers. We might as well just consider an ideal chain with no interactions between the monomers whatsoever. All we need to do is to imagine that this ideal chain has been squeezed into a cavity the sides of which are of length $R = \alpha a N^{1/2}$.

Take a certain monomer unit of the chain. Suppose it is currently far away from the cavity walls. What can we say about a strand of the chain near the chosen monomer? It does not seem to "know" anything about the surroundings. It can "sense" neither the walls of the cavity (since they are far away) nor the presence of other bits of the chain (since the chain is ideal). Therefore, this strand merely acts as a Gaussian coil. If its length is g, then its size will be about $ag^{1/2}$. Of course, this would only be right if $ag^{1/2} < R$, that is, $g < g^\star \sim (R/a)^2$.

Now let's look at a $g^\star$-strand. The monomers deep inside it can arrange themselves in any sort of shape. Therefore they do not contribute to the entropy (since they do not reduce the choice of possible conformations of the chain). On the other hand, the ends of the $g^\star$-strand must be near the walls, even though they cannot leave the cavity. This restricts the number of possible conformations, so that each end loses a bit of entropy, of order k. (To see this, suppose the number of conformations W of a monomer segment in (6.2) drops by half. Then the entropy decreases by $k \ln 2 \approx 0.76\, k$.) In a chain of N monomers, there are $N/g^\star$-strands altogether. So the loss of entropy is

$$U_{eff}(\alpha) = -TS(\alpha) \sim kT\frac{N}{g^\star} \sim kT\frac{Na^2}{R^2} \sim kT\frac{1}{\alpha^2}. \qquad (8.3)$$

For comparison, let us remember what we had for a swelling polymer, (i.e., for $\alpha > 1$) (see (7.9)):

$$U_{eff}(\alpha) \sim kT\alpha^2 \quad \text{when} \quad \alpha > 1. \qquad (8.4)$$

In (8.4) we left out a constant term (independent of α), since it does not affect any physically measurable quantities.

Now we know how the function $U_{eff}(\alpha)$ looks when $\alpha \simeq 1$ or $\alpha > 1$. Can we use this to figure out the form of $U_{eff}(\alpha)$ in the intermediate regime $\alpha \sim 1$? Since we only want a qualitative answer, we can just do a simple interpolation:

$$U_{eff}(\alpha) \sim kT\left(\alpha^2 + \alpha^{-2}\right). \qquad (8.5)$$

Estimate (8.5) gives the right result for $\alpha \ll 1$ and $\alpha \gg 1$, and is approximately correct (to an order of magnitude) at $\alpha \sim 1$. By the way, the function (8.5) has a minimum at $\alpha = 1$. This is not surprising. Indeed, if there are no excluded volume interactions, $U(\alpha) \equiv 0$, and so $F = U_{eff}(\alpha)$. The chain will behave as if ideal, which means that $\alpha = 1$.

8.5 ‒ The Swelling Coefficient α

Now we can summarize what we have learned. Based on (8.1), (8.2), and (8.5), we can write the free energy $F(\alpha)$ of a polymer molecule of size $R = \alpha N^{1/2}\ell$ taking into account excluded volume interactions:

$$F(\alpha) = kT\left(\alpha^2 + \alpha^{-2}\right) + kT\frac{BN^{1/2}}{\ell^3\alpha^3} + kT\frac{C}{\ell^6\alpha^6}. \tag{8.6}$$

Here, as usual, we omitted numerical coefficients of order 1 that should accompany each of the terms. This estimate gives a qualitatively correct answer for both $\alpha > 1$ (the swelling of a coil) and $\alpha < 1$ (the coil-globule transition).

What do we do next? Just as before, we need to minimize the free energy (8.6) as a function of α (see Section 7.5). The condition for a minimum, $\partial F(\alpha)/\partial \alpha = 0$, leads us to the equilibrium value of α. After substituting for the derivative, we have the equation

$$\alpha^5 - \alpha = x + y\alpha^{-3}. \tag{8.7}$$

Here $x \equiv K_1 BN^{1/2}/\ell^3$ and $y \equiv K_2 C/\ell^6$, where K_1 and K_2 are numerical constants of order 1. This equation determines the size of a coil, $R = \alpha N^{1/2}\ell$, as a function of two characteristic parameters, x and y.

A graphical interpretation of (8.7) is shown in Figure 8.3. We have plotted a set of curves $\alpha(x)$ for different values of y. How did we manage to do it? The trick is that you can easily find the reciprocal function $x(\alpha)$ from (8.7), for every given α and y. It is single-valued, and you can plot it on the graph of $\alpha(x)$.

What do we know about the parameters x and y? The sign of x merely matches the sign of the second virial coefficient B. In a good solvent, $B > 0$ and so $x > 0$. At the Θ point, $B = 0$ and so $x = 0$. And, finally, in a bad solvent (i.e., a precipitant), $B < 0$ and $x < 0$. What is the magnitude of x? In a very good solvent, the second virial coefficient B is neither especially large or small; in this case $x \gg 1$. This is because $x \sim N^{1/2}$ and $N \gg 1$. In a similar way, in a very bad solvent x is negative,

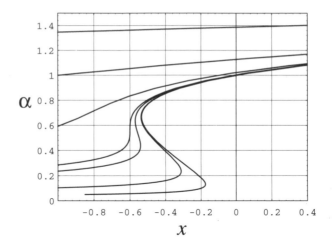

and $|x| \gg 1$. From all this we can conclude that as long as $|x| \sim 1$ we can be fairly sure that we are close to the Θ point. It looks like the parameter x always mirrors the quality of the solvent. As x ranges from $-\infty$ to $+\infty$, the quality of the solvent varies from very bad ($|x| \gg 1$, $x < 0$) to moderately bad ($|x| \sim 1$, $x < 0$), to the Θ solvent ($x = 0$), to a moderately good solvent ($|x| \sim 1$, $x > 0$), and finally to a very good one ($|x| \gg 1$, $x > 0$). According to Section 7.4, the quality of the solvent is controlled, in particular, by the temperature. Thus we can relate the parameter x to the temperature. The range $x < 0$ corresponds to temperatures $T < \Theta$, and $x > 0$ corresponds to $T > \Theta$; x is a monotonically increasing function of temperature. This is why the curves $\alpha(x)$ plotted in Figure 8.3 reflect the dependence of a polymer's swelling coefficient on the temperature or on the quality of the solvent.

As you can see from Figure 8.3, the main changes in the function $\alpha(x)$ occur in the vicinity of the Θ temperature. This is the region where the parameter $y = K_2 C/\ell^6$ hardly varies, so we can set it to some constant value (as in fact we did in Figure 8.3). By definition, y depends on the third virial coefficient C, that is, on the makeup of the polymer chain. We shall skip the details of the theory and just tell you some results. The study of the coefficient C for different types of monomers has shown the following. If a polymer chain is flexible (which means that the Kuhn segment ℓ is of the same order as the characteristic thickness of the chain d), then $C \sim \ell^6$, so that $y \sim 1$. For rigid chains ($d \simeq \ell$), the stiffer the chain, the smaller the parameter y. Thus, all the different curves in Figure 8.3 correspond to different amounts of rigidity of the chain. The parameter y describes the rigidity.

You may have noticed that all the curves $\alpha(x)$ in Figure 8.3 fall into two essentially different groups, depending on the value of y. If $y > y_{cr} = 1/60$ (i.e., if the chain is fairly flexible), α grows monotonically with x, although the rate of this growth varies. It changes slowly in the range $x \simeq -1$, and also when $x > 0$, but goes up rather rapidly for $x \sim -1$ (i.e., just below the Θ temperature). On the other hand, if $y < y_{cr}$, there is a typical "loop" just below the Θ temperature. It looks like the loops on isotherms of a van der Waals gas. So the function $\alpha(x)$ becomes multivalued: There are three values of α for each value of x. This gives the function $F(\alpha)$ in (8.9) as many as three extrema (Figure 8.4), two local minima and one maximum.

The two local minima we are interested in correspond to the smallest and the largest of the three values of α for each given x. Now we need to compare the values of F at the two minima, to discern which is smallest. It will give us, as usual, the equilibrium value of the swelling coefficient α. This is illustrated in Figure 8.5. Let us look at one of the curves. You can see that, as long as x is less than some critical value x_{cr}, one of the two branches of the "loop" provides the equilibrium solution for α. However, as x increases, there is a sudden "swapping over" between the branches. Now the equilibrium solution is represented by the other branch. Thus, if $y < y_{cr}$ (which means that the chain is fairly stiff), the polymer suddenly rearranges its own shape just below the Θ point. When it does this, it changes in size in a very abrupt, jump-like manner. The smaller the parameter y the more dramatic this "jump." (see Figure 8.3).

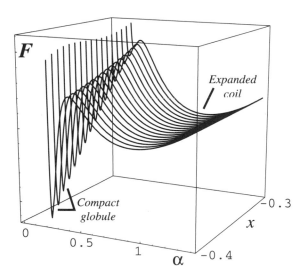

FIGURE 8.4
The dependence $F(\alpha)$ in the case where $\alpha(x)$ is multivalued. As x changes (which can be controlled by, say, temperature change), the shape of the $F(\alpha)$ dependence changes such that one minimum gets deeper at the expense of the other. Deeper minimum corresponds to the more stable state. For this figure, we choose the value $y = 0.001$.

FIGURE 8.5
The curves $\alpha(x)$ in Figure 8.3 are multivalued at some x; in this figure, one solution is selected for each x such that the values of $\alpha(x)$ correspond to the absolute minimum free energy for every x. The values of y are the same as in Figure 8.3.

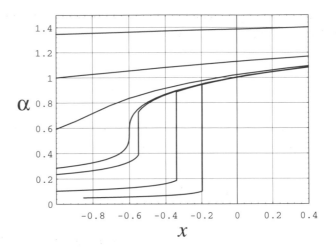

8.6 — The Coil-Globule Transition

Now let's explore the dependence $\alpha(x)$, given by (8.7), in more detail in the range $y < y_{cr}$. When $x > 0$ (so that the solvent is good), this dependence provides the correct qualitative description of how a polymer coil swells due to excluded volume interactions (see Section 7.5). In this case $\alpha \gg 1$, so we can neglect the second terms on both sides of equation (8.7). Then we shall have

$$\alpha \sim \left(\frac{BN^{1/2}}{\ell^3}\right)^{1/5}, \quad \text{i.e.,} \quad R \sim \alpha\ell N^{1/2} \sim N^{3/5}\left(\frac{B}{\ell^3}\right)^{1/5}\ell, \quad (8.8)$$

in complete agreement with (7.13).

Now, what if $x < 0$? Let's look at the region $x > x_{cr}$ first, which corresponds to higher temperatures than that where the jump in polymer size occurs. We can deduce from (8.7) that α is close to one in this case. This means that the molecule takes the shape of a nearly Gaussian coil and is hardly disturbed at all by excluded volume interactions. What happens if $x < x_{cr}$ (on the bottom branch of the "loop")? In this case, normally, the equilibrium swelling coefficient is very low ($\alpha \simeq 1$). So the molecule looks terribly "squashed" by attraction between the monomers, when compared with an ideal coil. The terms on the left-hand side of equation (8.7) will be much smaller than those on the right. Where did the terms on the left come from? You can easily trace that they have to do with the entropy (see (8.2) and (8.5)).

From here we conclude that the entropy contribution to the free energy $U_{eff}(\alpha)$ is not significant for $x < x_{cr}$. Thus, the equilibrium size of the molecule is only controlled by the free energy of the monomer interactions $U(\alpha)$. If we neglect the terms α^5 and α in (8.7), we shall have

$$\alpha \sim \frac{C^{1/3}}{(-B)^{1/3} N^{1/6} \ell}. \tag{8.9}$$

(Since $x < x_{cr} < 0$, the second virial coefficient stands for attraction and so must be less than zero.) We can rewrite it for the equilibrium size R:

$$R \sim \alpha N^{1/2} \ell \sim \left[\frac{C}{(-B)} \right]^{1/3} N^{1/3}. \tag{8.10}$$

Then the concentration n of monomers inside the molecule is estimated as

$$n \sim \frac{N}{R^3} \sim \frac{(-B)}{C}. \tag{8.11}$$

According to equation (8.11), the shape of the molecule for $x < x_{cr}$ is totally different from that of a typical polymer coil (e. g., Figure 2.6). First of all, the monomer density does not fall as N grows (compare (8.11) with (7.4)). Moreover, the size of the molecule is proportional to $N^{1/3}$ (not to $N^{1/2}$ as for an ideal polymer, or $N^{3/5}$ as for a polymer with excluded volume interactions). Such unusual properties are actually similar to those of an ordinary liquid drop of constant density. This suggests that if $x < x_{cr}$ the molecule might have a globular structure, just like the one in Figure 8.1. This turns out to be true.

Thus, it is the coil-globule transition that is reflected by the "jump" in the molecule's size in Figure 8.5 for $y < y_{cr}$. You can see that this transition occurs at $x_{cr} \sim -1$, which is only slightly below the Θ temperature. What does this mean? Suppose you have a loose polymer coil at the Θ point. Then it does not take much to make the coil "condense onto itself" and form a globule. Just a slight worsening of the solvent quality—that is, just a tiny bit of attraction between the monomers—would be enough.

So far, we have only considered the coil-globule transition for $y < y_{cr}$. This is when it is accompanied by a "jump" in the molecule's size. Can the coil-globule transition also happen in the case of $y > y_{cr}$, that is, when the polymer chains are fairly flexible? Certainly it can. If the temperature falls far below the Θ point, attraction between the monomers becomes strong enough in this case too. As a result, a condensed globular state is formed. Mathematically, we can describe it in the same fashion as before, skipping the terms on the left-hand side of equation (8.7). We end up with the same estimates (8.9–8.11). Still, there is an important difference from the previous case $y < y_{cr}$. Now the formation of the globule is not steplike, but is a smooth and gradual process. However, although it happens smoothly, it only spans over a fairly narrow temperature interval, somewhat below the Θ point. You can see this in Figure 8.3. The transition region for $y > y_{cr}$ is just where the function $\alpha(x)$ changes most rapidly. All this region lies within $|x| \sim 1$, which corresponds to a temperature variation of the order

$$\frac{(\Theta - T)}{\Theta} \sim N^{-1/2} \ll 1. \tag{8.12}$$

If the relative temperature variations are much greater than $N^{-1/2}$, you can be absolutely sure that the molecule is in the globular state, even for $y > y_{cr}$.

8.7 — Pretransitional Swelling

There is an extremely important feature of the coil-globule transition that we can draw out of equation (8.3). We have already said that the transition occurs in the vicinity of the Θ temperature. However, at the Θ point the second virial coefficient B goes to zero and is very small around that temperature. What does this tell us? Suppose we approach the Θ temperature from below (i.e., we move toward the transition region). We shall notice that the globule grows significantly in size. Meanwhile, the average concentration n of the monomers inside the globule decreases just as significantly. In other words, the globule gradually swells. (This is certainly because B decreases and tends to zero as $T \to \Theta$.)

This is good news. It means that we need to keep no more than the two first terms in the virial expansion (see (8.2)). That will do for globules near the Θ temperature (i.e., near the coil-globule transition point). Indeed, since the monomer concentration in the globule is fairly low, we can ignore many-body interactions. Thus, all we need to describe a globule near the Θ temperature are the second and the third virial coefficients, B and C.

Now you see that the analogy between the coil-globule transition and an ordinary gas-liquid transition has its limits. A liquid condensed from a gas always has a fairly high density. In contrast, a "newly born" globule is usually quite tenuous at the transition point. This explains why a theory for the coil-globule transition is much more straightforward than for a gas-liquid one. In the case of the globule, we are equipped with a nice small parameter, the monomer concentration inside the molecule. This really helps when constructing a strict mathematical description.

Thus, when nearing the Θ point, the globule swells. This gives rise to fluctuations. An interesting question is this. At what stage will the growth in fluctuations cause the actual transition? In fact, when the globule is swelling it remains a globule until it gets really very close to the Θ temperature. In other words, the correlation radius of the concentration fluctuations remains much shorter than the size of the molecule. The actual transition to the coil (i.e., to much stronger fluctuations) only occurs at the temperature T estimated by (8.12). If the chain is fairly flexible (i.e., $y > y_{cr}$), the transition is smooth. The globule goes on swelling more and more, until it reaches the size of a normal polymer coil at the Θ point. In contrast, if the chain is stiff (i.e., $y > y_{cr}$), the transition has the form of a jump, which happens at some critical value x_{cr} (i.e., at the critical temperature T_{cr}).

8.8 — Experimental Observation of the Coil-Globule Transition

There have been a fair number of experiments on the coil-globule transition with decreasing temperature. The most detailed study was carried out by scientists from three laboratories. One was at the Institute of High Molecular Weight Compounds in St. Petersburg where they used a polarized luminescence technique. Another was at the Massachusetts Institute of Technology, by means of inelastic scattering of laser beams by polymer solutions. Recently, the main contribution has been made by the group at SUNY Stony Brook. We shall not go into the details of how exactly the SUNY experiments were done. In general, they measured some quantities related to the diffusion of the molecules (both translational and rotational), from which the average size R of a molecule can be deduced.

Concerning the system examined, most of the experiments were done on polystyrene macromolecules diluted in cyclohexane. The Θ temperature for this system is within the convenient range; it is close to $35°C$. What was noticed in the experiments? Below the Θ point, the molecule exhibits a very abrupt shrinkage in an interval of only a few degrees. Its volume changes by ten times or more. Obviously, this is where the molecule turns into a globule. However, at the very point of the coil-globule transition, the monomer density inside the globule is still much less than for a dry polymer. In other words, the globule is still rather loose, just as the theory predicts.

Some other methods have also been employed to observe the coil-globule transition for isolated polymer molecules (e. g., ordinary light scattering, viscosity and osmotic pressure measurements, and elastic neutron scattering off polymer solutions). However, the two techniques we mentioned before are the best for this purpose. They are very sensitive and allow measurements of solutions at extremely low concentrations.

The reason the low concentrations are so crucial is this. Suppose we reduce the temperature below the Θ point. Attraction between the monomers starts to prevail. This certainly enhances the "condensation" within the molecule (i.e., the formation of a dense globule). This is not the only tendency, however. Another is for different molecules to stick together to form huge "lumps." or aggregates, and these molecular aggregates precipitate out. Obviously, we want to avoid such a process. Therefore, we need to make sure that the condensation inside the molecule is much more likely than that outside (at the transition point). The only way to do this is to restrict ourselves to low concentrations. The less concentrated the solution we choose, the further below the Θ temperature we can go without worrying about molecular aggregates. In reality, the experiments go to a concentration as low as $c = 10^{-2}$ g/l for chains that are as long as about 10^7 monomer units each. Nevertheless, it remains unclear whether this concentration is low enough to study the entire region of the coil-globule transition.

Despite various tricks, the problem of chain aggregation has not quite been solved. The experts still argue. Research continues, and experimentalists have now reached remarkably low concentrations. Just imagine: it takes almost 10 minutes between two consecutive collisions. If we can we manage to do all the measurements in these 10 minutes, we would not care whether the molecules precipitate out afterward! This seems to be the most promising direction for further studies.

There is a special movie, "*Collapse*," on the CD ROM showing polymer collapse. Look at how compact the chain gets in the end. This is obviously because monomers attract each other. Also look carefully at how the process develops: It reminds one indeed of our Figure 8.6, doesn't it? Of course, the simulated chain is relatively short, this is why there was no room to develop all the levels of self-similarity as beautifully as shown in Figure 8.6. Nevertheless, it can be clearly seen how the chain gets gradually shorter and thicker. Of course, the chain itself does not change, but its conformation transforms such that it looks shorter and thicker. It was P. G. de Gennes who first predicted this scenario of collapse, and people call it the "sausage model." Well, some people argue that in fact it is not really a sausage, because it is not evenly thick all along its length. Rather, because of the beads, these people want to call it the "necklace model." Certainly, a necklace is more beautiful than a sausage, but as far as science is concerned, we feel that the difference is not so important (though we do recognize the difference between a necklace and a sausage in real life!).

8.9 — Dynamics of the Coil-Globule Transition

Whatever transition you explore, it is not just the initial and the final states that you are interested in, but the actual process of the transition. You are not only concerned with water, steam, and so on, but also with boiling, vaporization, and condensation; nor only with ice, but also with melting and solidification.

What can we find out about "globulization"? How does it proceed? How do polymer networks collapse? Figure 8.6 sketches the first stages of globule formation. At the very start, lots of little "droplets" emerge. They are the "embryos" of the globular phase. Then the "embryos" grow and merge with each other, until a larger spherical globule is formed.

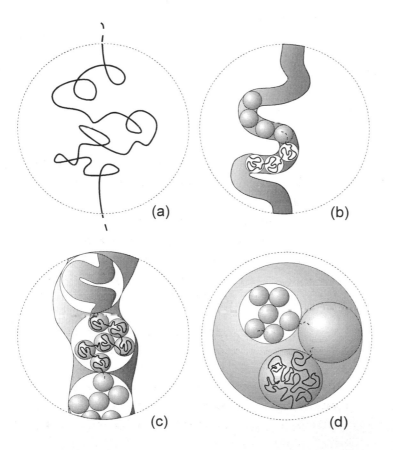

(a)   (b)   (c)   (d)

FIGURE 8.6
A few initial stages of the coil-globule transition. This looks self-similar! (Compare with what we write about self-similarity in Chapter 10). *Source:* Courtesy of S. Nechaev.

But theorists think that this may not yet be the equilibrium globule, because its chain has not yet become entangled. To form the entanglements, it needs to go through an extra stage, which has to do with the so-called reptational motion. We shall discuss reptations a little later, in Chapter 9. The interesting thing is that there is a very simple experiment that you can do yourself to check this theoretical idea. Take a piece of rope (a "macromolecule"), crumple it (the "coil-globule transition"), and try to disentangle it without much shaking ("thermal motion"). People like mountain climbers, who are familiar with ropes, know that this is quite straightforward, as long as you don't pull the ends! This suggests that mere crumpling (i.e., collapse) is not enough to create the knots.

It sounds quite a nice theoretical idea, that the chain has to be crumpled first and then entangled. Recent subtle experiments by B. Chu offer some indirect support for this view. However, if you believe it, you will have to face another bunch of questions. Imagine a very entangled globule. What if we tried to transform it back into a swollen coil? Using the same analogy with a rope, we can anticipate that the knots will tighten up, and.... Well, we all know how long it often takes to undo a tight knot! Speaking of the molecular chain, would thermal motion ever be able to undo the knots? Would it take only a microscopic instance of time, or a minute? Or maybe a couple of hours? Many years? Eternity? We do not know yet and the uncertainty is tantalising!

8.10 — Some Generalizations

Let us summarize what we have found out about the coil-globule transition in an isolated homopolymer molecule. Its experimental and theoretical investigation is certainly very important. It is the simplest of the intermolecular condensation phenomena. If you understand it, you will be able to move on to more complex effects. On the other hand, it can only be observed under very special conditions (like low concentrations of polymer in solution). Fortunately, it turns out that there are many other transitions of a very similar sort in the physics of polymers and biopolymers. Not all of them are so capricious; quite often the problem of precipitation is nonexistent or not that crucial.

We compared the coil-globule transition with a gas-liquid one. For more complex polymer systems, you can also think of some analogies. The local microstructure of a globule may sometimes resemble a liquid or plastic crystal, an amorphous solid, a glass, an ordinary crystal, a solid or liquid solution, and so on. This is why such strange things as a globule-globule phase transition become possible. (What happens is that the globule's core just rearranges its structure.)

Another interesting thing is liquid-crystalline ordering in a concentrated solution of rigid polymer chains. (This is when the chains have a predominant orientation; see Section 3.5.) It turns out that this ordering can also be regarded as the formation of a globule! You only have to imagine it in a special sort of space—the space of the segments' orientations, not in the usual three-dimensional space.

In the next few sections, we shall look at three different effects that have something in common. In a way, they are all similar to the coil-globule transition. These effects are the collapse of polymer networks, the formation of a compact DNA, and denaturation of proteins. Of course, there are many more such phenomena. If you are intrigued, you can find lots of details, for example, in our book *Statistical Physics of Macromolecules* [4].

8.11 — The Collapse of Polymer Networks

Suppose we have a piece of polymer network, swollen because it is in a good solvent. Let's look at one of the subchains (that is, a part of a chain confined between two adjacent cross-links; see Section 6.5). Naturally, it tends to take the shape of a loose polymer coil typical of a good solvent. Now, say the solvent becomes worse. The subchains will shrink, which leads to the shrinking of the whole network. If the temperature drops below the Θ point, each of the subchains will undergo a coil-globule transition. As a result, the entire network will rapidly collapse.

This is exactly the process called the collapse of polymer networks. It was discovered by T. Tanaka and his colleagues at the Massachusetts Institute of Technology in 1978. They used networks of polyacryl amide diluted in a mixture of acetone and water. In these experiments, the temperature was not varied. To make the solvent worse, they just poured some extra acetone into the solution. (This worked because acetone, in contrast to water, is a bad solvent for polyacryl amide.) Figure 8.7 gives an idea of what was found. It sketches how the size of the network depends on the acetone concentration. You can see that if you "dump" 42% of acetone, the network collapses suddenly. Its volume drops by a factor of nearly 20.

When we are dealing with the collapse of networks, there is no such problem as the precipitation of molecular aggregates. We have only one macroscopic sample that shrinks as a whole when all the subchains turn into globules.

It seems that to develop a theory for the collapse of networks, we could follow the same logic as for the coil-globule transitions. After all, they both have the same cause. However, this does not work. Such a theory was indeed proposed, but

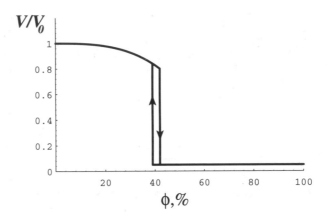

experiments did not support it. You can even spot a contradiction in Figure 8.7. The data are for polyacryl amide, which is a flexible polymer. So we are in the regime with $y > y_{cr}$, and the coil-globule transition should happen smoothly. Yet, the curve in Figure 8.7 drops down in a step.

Furthermore, Tanaka carried out a detailed investigation and found the following. The height of the step depended strongly on the time interval between preparing the network and starting the experiment. The longer the delay, the higher the step. If a network was kept for two months after it was made, changes in volume by factors of a few hundred were observed. In contrast, freshly prepared networks manifested a nice, smooth collapse.

Can we find an explanation for all these oddities? The clue is that polyacryl amide chains are not stable in water. They are prone to the chemical reaction called hydrolysis. As a result, monomers that are initially neutral dissociate. This means that small light ions split off from the monomers, leaving behind segments of the opposite charge. (Such small ions are usually called counterions, as we discussed in Section 2.5). The estranged counterions float on their own inside the swelling network. The hydrolysis of polyacryl amide occurs extremely slowly. So, over a short period of time, only a very small proportion of the monomers will gain an electric charge. However, the "older" the network (i.e., the longer ago it was prepared), the higher will grow the proportion of charged monomers. Now we can explain Tanaka's experiments, if we make just one assumption. We need to assume that even a small proportion of charged monomers can affect the collapse rather strongly and that the step becomes even higher as this proportion increases.

What gives us the right to make such an assumption? Let us see. Together with the charged monomers, there are floating counterions in the swollen network (Figure 8.8). Note that the counterions do not drift out of the network into the

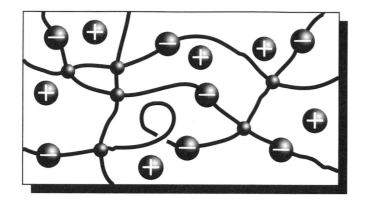

FIGURE 8.8
The counterions in a charged polymer network.

pure solvent. Why not? If they did, the system would lose its electrical neutrality on a large, macroscopic scale. Strong Coulomb interactions would arise between the charges on the network and the counterions outside. The energy of these interactions is extremely high. This is why such a state is energetically unfavorable and never occurs.

Thus, the counterions move freely inside the network, but are not allowed outside. You could say that the "shell" of the network (i.e., its outside surface) stops them. Evidently, crowds of counterions exert some pressure on the "shell." This pressure stretches the network more in all directions. We are going to show now that this is exactly what makes the collapse so different from what you might expect.

The free energy of the whole network is the sum of the free energies of all the individual subchains. The free energy of each subchain, in its turn, consists of entropy and energy terms, $U_{eff}(\alpha)$ and $U(\alpha)$ (equation (8.1)). (Here α stands as usual for the swelling coefficient of the entire network.) The dependences $U_{eff}(\alpha)$ and $U(\alpha)$, calculated from equations (8.2) and (8.8), are plotted in Figure 8.9a, together with their sum $F(\alpha)$, for an electrically neutral polymer network. The chosen value of the parameter $x = K_1 BN^{1/2}/\ell^3$ lies in the coil-globule transition region. At the same time, the value of $y = K_2 C/\ell^6$ is set to just over y_{cr}, so the function $F(\alpha)$ has only one minimum (this corresponds to the "no loop" case in Figure 8.3).

What will change if we create just a small proportion of charged monomers in the network and the same number of counterions? The energy $U(\alpha)$ will hardly alter. There will certainly be some extra contribution to it, due to Coulomb interactions, but it will not really matter given that the number of charges is small. As for $U_{eff}(\alpha)$, we must not forget what we have just discussed. The counterions cause osmotic pressure, which inflates the network. What difference

FIGURE 8.9
(a) The dependences
$F(\alpha)$, $U_{eff}(\alpha)$, and $U(\alpha)$
for a neutral network.
(b) The change in
$U_{eff}(\alpha)$ when the
network acquires an
electrical charge. (c)
$F(\alpha)$ for a charged
network.

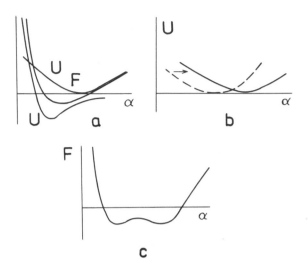

does this make? First of all, the minimum in $U_{eff}(\alpha)$ is shifted toward higher α
(Figure 8.9b). As a result, the function $F(\alpha)$ may now appear as in Figure 8.9c.
This corresponds to the "loop" in Figure 8.3 and implies a steplike collapse.

Thus, the charged monomers in the network extend the range of parameters
for which a steplike collapse may occur. Even networks of flexible chains ($y > y_{cr}$)
which are normally expected to have a smooth collapse (when they are neutral,
Figure 8.9a), may exhibit a step (Figure 8.9c). Calculations show that if you have
just a few charges per subchain, you can nearly guarantee that the transition will
be steplike. Now you can see why partially charged networks, although made
from flexible polyacryl amide chains, collapse so abruptly (Figure 8.7). Look at
Figure 8.9 again. The more charged monomers you have (i.e., the higher the
inflating osmotic pressure from the counterions), the higher the step. This fully
explains the observations.

The collapse of polymer networks has recently attracted a lot of attention. This
boom is partially due to some important applications, which all stem from the
fact that only a slight change in the quality of the solvent will make the network
collapse rapidly. It is especially useful that the collapse is very sensitive to the
presence of charged monomers and counterions in the solution. Thus collapsing
networks can be adapted to detect small ion impurities in a solution, as well as to
clear the impurities away. Besides all this, the collapse of networks can also serve
as a good model for some other processes in biology (e. g., in the glassy medium
in the eye).

Tanaka's group at MIT have "collected" a great variety of different cases.
For example, the collapse can occur if the temperature changes (to complicate

matters, some networks collapse on cooling and some on heating). It can also be caused by ion forces, by adding certain molecules, by light, and so on. This is pictorially illustrated in Plate 6, drawn by Tanaka himself. Moreover, a network can be made to collapse in patches, to give an irregular density. In this case we would obtain a special kind of diffraction grating, or a hologram of the object. (Who knows what could make up the memory of future computers?)

We should not forget about gel swelling. As you should be ready to appreciate, all the subchains of the gel will swell once placed in a good solvent, and this will make the macroscopic gel sample swell. What is interesting about it is that good solvent diffuses rather slowly into the volume of the gel, and thus surface portion of the gel swells faster than the balk. This leads to the formation of beautiful and sometimes strange looking patterns, like the one shown in Plate 7.

To conclude, we will mention just one more peculiarity of polyelectrolytic gels. It can be easily understood if we use some things that we discovered before. We saw that the counterions of polyelectrolytic gels cause an extra pressure that makes the network swell. This pressure is really rather high. So the gel can swell quite dramatically, to the extent that there will be no more than 0.1% of polymer inside the gel, and the rest will be taken up by water. In other words, 1 g of a "dry" polymer gel can absorb up to 1 kg of water! This is why they say sometimes that polyelectrolytic gels are superabsorbers of water. This property has found many applications. Perhaps the most impressive one is babies' nappies (disposable diapers). How do they manage to take in so much liquid? What happens is that the water is absorbed by granules of polyelectrolytic gels made of polyacrylic and polymethacrylic acids. Such gels are also used in agriculture, to keep the upper layers of soil humid in dry areas.

8.12 — The Globular State of the DNA Double Helix

In real life, in cells and viruses there is so little room to store genetic information that a usual coil of DNA double helix just would not fit. You can see this in Plate 2: Even when the coat of bacteria is almost completely destroyed, DNA remains globular (compare with Figure 8.1). Hence, DNA has to be stored in a compact globular shape. In fact, it makes a very complex globule. It not only contains a long DNA string, but also carries various smaller proteins attached to the string in a special manner. For example, in viruses, DNA is covered in a kind of protein "coat." In cell nuclei, certain proteins (histones) exist that play a number of roles. They form the so-called nucleosomes, which act as "reels" for fairly short strands of DNA (about 1000 "letters"). Another complication is that natural DNA molecules

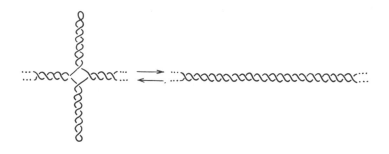

FIGURE 8.10
The cross-like structure of a palindromic strand of DNA. To make this structure possible, the sequence must be symmetric, or be a palindrome. The second string's palindrome is not identical, but complementary to the first string's palindrome.

are normally (probably always?) closed and have lots of knots and entanglements with each other (Figure 2.9*e* and Plate 8). And lastly, some DNA sections, with particular types of primary structure, may form very unusual secondary structures under certain conditions that can occur in a globule. These structures are different from the familiar Watson–Crick right-handed double helix (which is called the *B*-form). In particular, there is a left-handed double helix (the *Z*-form), a triple helix (the *H*-form), and so on. For example, you can encounter palindromes in the primary structure of segments of DNA. (Palindromes are sentences that read the same in both directions, e.g., "A man, a plan, a canal—Panama," "Draw pupil's lip upward!," "And DNA," etc.) Palindromic bits of DNA often take the shape of a cross (Figure 8.10).

These examples show that it is very hard to study natural DNA globules. So it makes sense to start with DNA placed in a bad solvent to see what we can find out about the coil-globule transition in this case. It is actually quite interesting.

There is a very simple movie called *"Knot"* on the CD ROM. It shows a polymer chain that is knotted from the very beginning, and no matter what happens to this chain later in the thermal bath of the solvent, it cannot unravel the knot. Then another polymer with a knot, which at first looks the same, is shown. Because it is an open polymer with free ends, it eventually gets rid of the knot and swells.

The movie *"Link"* is fairly similar to *"Knot."* Here we have two polymer rings that are entangled with each other, and they cannot change their entanglement. This type of link, though of much higher order, often exists between two strands of DNA, since in living cells DNA usually has a closed circular form.

First of all, DNA globules have the shape of a torus (like a doughnut). This is not surprising. There are no places in the double helix where it can be easily kinked. (In other words, DNA behaves like a wormlike chain; see Section 3.3.) This is why it cannot possibly fill in the core of the globule, and we end up with a

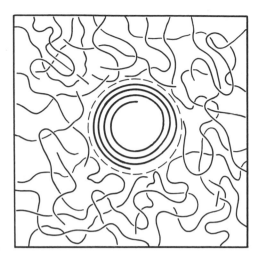

FIGURE 8.11
Formation of a DNA globule in a polymer solution.

hole. Another interesting thing is that the double helix is very stiff. Its persistent length (50 nm) is much greater than its width (2 nm). Hence, the Kuhn segments packed into the globule tend to line up parallel to each other, in every region of the globule. Thus, they create a liquid crystal inside the molecule! Also, it is quite a tricky task to find a solvent bad enough to turn the stiff double helix into a globule. The best, it seems, is to take a solution of another polymer, whose coils are comparatively short and flexible, so that they will be expelled when DNA shrinks into a globule and will not get in the way of its inner structure (Figure 8.11). The "gas" of such coils causes a kind of external pressure on the "walls" of the DNA globule.

8.13 — Globular Structure of Proteins and Conformational Transitions in Globular Proteins

There are many globular proteins in a living cell, and they play a key role. We have already discussed this in Chapter 4. However, the theory of such systems is extremely hard; a protein globule is perhaps one of the most complex objects in modern physics. What is most striking and unusual is that proteins have a strictly defined spatial tertiary structure (see Sections 4.6 and 4.7). Of course, it is nothing like an ordinary globule, although we know that the fluctuations in an ordinary globule are very small too. However, the fluctuations are only small for such general, overall quantities as the size of the globule or the monomer density. In contrast, in a protein globule, the entire spatial structure of the whole chain is fixed. So in this case we are talking about small fluctuations not of some vague quantities, but of the actual coordinates of most atoms!

You may wonder why we are so worried about the tertiary structure in particular. There are other things, like the primary structure, for instance, which means that the whole sequence of monomers is also fixed! It all has to do with how different structures are formed. To produce protein chains with the right primary structure (as determined by the genetic program, DNA), there is a special complex "machine" in the cell, called a ribosome. Unfortunately, we do not yet know how to synthesize specific proteins without cells. Scientists certainly hope that they will eventually figure out how things work in a cell, but, at present they cannot do much better than brush the question aside, saying; "Well, there is some mechanism of biosynthesis . . . . "

How is the fixed tertiary structure created? Perhaps there is some other mysterious "machine," of which we know nothing at all, but which is actually in charge of packing protein chains into globules of the right shape. Note: this is where we reach the crucial point. The answer is that such a "machine" does not have to exist. We can pretty well manage without it. This was first shown by the American biophysicist Christian Anfinsen in 1962 (he later received a Nobel prize for this work).

Anfinsen's idea was the following. We have already mentioned denaturation of proteins; it can be caused by heating or by adding some low molecular weight substances to the solution. Denaturation is a sharp conformational transition. The stiff globular structure is destroyed, and the protein ceases to be chemically active. Anfinsen wondered if the protein could be returned to the natural state. He tried and succeeded. Thus, we do not have to employ a living cell or borrow some special living "machine." The correct tertiary structure can be restored *in vitro* (renaturation). All we need is to be careful: make sure that the process is very gradual, that the concentration is low, and so on.

As a matter of fact, as usual in biology, every rule seems to have at least some exceptions. Some complex proteins fold with the help of special molecules called chaperons. Nevertheless, a firmly established fact is that many proteins do not need any assistance and are able to do this amazing job easily and reliably. Moreover, from the physics point of view, does it really make a great difference whether it is just one protein molecule that organizes itself, or a pair of molecules, such as a protein and a chaperon?

Thus, in contrast to the primary structure, which can only be produced in the living "factory," the tertiary structure is capable of organizing itself. This capability of self-organization is exactly what enables all proteins to function, and all life to live.

At first glance, self-organization seems very straightforward. Suppose we have a certain chain with a fixed sequence of monomers. In the right sort of circumstances, left on its own, it will always roll up into a coil in exactly the same way. This may not seem strange, yet this phenomenon is unique. It is completely unlike anything else in physics. To try to understand it has been a challenge for physicists for over a quarter of a century. Let's look at this problem in more detail.

Look at Plate 9: If a sequence of units of a protein chain is similar to a sequence of letters in a certain text or message, then we immediately discover that we cannot read it! It is just that we do not understand the language. It turns out that the meaning of the message can be revealed when the molecule collapses. You could say that the tertiary structure contains the meaning of the primary text. Thus, in contrast to a simple collapse of a homopolymer, a self-organized collapse of a protein chain can be described as reading *with understanding*.

In a sense, a protein globule has something in common with a solid crystal. They both have a very well-defined three-dimensional structure, which is due to the small magnitude of fluctuations. However, this analogy does not extend any further than just the rigidity of the spatial arrangement. There is no similarity between the two structures themselves. Crystals are distinguished for their periodicity. In contrast, a protein globule consists of assorted amino acid residues, and so is completely nonuniform and irregular in shape. In this sense, a protein globule resembles so-called disordered systems, also known to physics (such as a glass). However, we shall see shortly that this analogy is very limited too. It does not go any deeper than the very fact of disorder.

What do we really mean by "a glass"? In Chapter 3 we talked about both polymer and nonpolymer glasses. We can describe them as substances that have been "frozen" in a nonequilibrium state. Glasses have enormous relaxation times (i.e., the time for the system to reach equilibrium). This takes far longer than any sensible physical experiment you can think of. We could say that the glass has "memorized" the structure (i.e., the positions of the atoms) that it happened to have when it was made (i.e., as it was cooled). If we melt the glass and then cool it down again, a new microstructure is created, so the previous "memory" will be totally lost. Equilibrium is never reached. That is exactly why glasses are disordered.

In fact, there is a remarkable resemblance between a glass and the primary structure of biopolymers. (Unfortunately, you cannot exploit this very much.) More or less spontaneous rearrangements of the primary structure take unthinkably long times. Thus, we can also talk of "memory." As soon as the primary structure is destroyed (i.e., the chain is broken), the "memory" is completely lost. However, the tertiary structure is very different. It does not have that much in common with a glass, except merely the lack of order. The tertiary structure is never "forgotten," even after denaturation. If it were, it would never be able to reorganize itself. Of course, all the "memories" are kept in the primary structure, engraved in some unknown secret language.

Thus, the tertiary structure can rebuild itself just like a stable crystal, and it is irregular in shape just like a nonequilibrium glass.

What causes the irregularity? Is it just because the monomers are rather unlike each other and do not fit into a nice shape? Suppose this were true. Take a particular collection of such "awkward" monomers (i.e., fix the primary structure). You could think that the spatial arrangement they choose (i.e., the tertiary structure) will correspond to a simple equilibrium. Then the self-organization seems no surprise. It will be the same sort of process as when a ball rolls down into a hole.

Does this sound convincing? Not really. How does the chain manage to find the equilibrium? Let's make an estimate. Say the chain is as short as 100 units. Each bond of the chain can take two different conformations only, for example, "right" and "left." (This is certainly an underestimate!) Even then the chain can have as many as $2^{100} = (2^{10})^{10} = (1024)^{10} \approx (1000)^{10} = 10^{30}$ conformations in total. Now, suppose the conformation is changed by every single atomic collision, that is, every 10^{-11} s. How long would the chain take to go through all the conformations in search of the stable state? Our calculation gives the incredibly long time of $10^{30} \cdot 10^{-11}$ s $= 10^{19}$ s $\approx 10^{12}$ yr. (Just for comparison, the age of the Universe is guessed to be about 10^{10} years!). This problem is known as Levinthal's paradox.

Alas, we seem to be getting nowhere. We had hoped to learn about self-organization through some analogies with other physical objects. However, we have found none. What are we left with? Let's see what the physics of self-organization has achieved so far.

Self-organization can be regarded as renaturation, the process opposite to denaturation of proteins. This is why a great deal of effort has been put into the study of denaturation. The first way in which denaturation may be caused is by heating. This is accompanied by absorption of heat. (Roughly, as much heat is absorbed as when a piece of crystal of the same size is melted.) Another way to cause denaturation is to add a special substance to the water, to reduce the hydrophobic effect (see Sections 4.1 and 4.7). Alternatively, some alkali or acid may be added. The explanation is simple. Certain amino acids gain a positive electric charge in an acidic medium, and others gain a negative charge in an alkaline medium. In both cases globules will become unstable, due to repulsion between similarly charged segments.

Some biologists were rather sceptical about studying denaturation. They thought it was a bit like "searching under a lamppost." You do not explore what is interesting (i.e., how proteins work), but what is easy. But some physicists (in particular, I. M. Lifshitz and O. B. Ptitsyn in Russia) formed the opposite camp. They reckoned that any conformational transition is worth investigating. It gives us more of an idea of the whole scope of possible states. Furthermore, the biologist V. Ya. Alexandrov discovered an interesting correlation. Suppose you have a little protein taken from a living creature. It turns out that the temperature of the protein's denaturation is related to the normal temperature of the creature's body. For example, the proteins from bacteria that live in glaciers were much less heat-resistant than those from inhabitants of water near warm geysers.

Eventually, in 1982, the physics of denaturation had its first serious success. O. B. Ptitsyn and his colleagues discovered that, in many cases, denaturation is not a globule-coil transition. It is rather a transition from the natural globule to the molten globule. During this transition, the blocks of the secondary structure remain stiff, but open up a little. This gives more room to the amino acids' side groups, which can then oscillate and rotate more freely. However, the globule as a whole remains stable. The openings are too narrow to let in water molecules, so the hydrophobic effect is not disturbed (Figure 8.12).

The image of the molten globule helps to explain experiments on the dynamics of self-organization. There are two stages to this process. The first one is reasonably quick. It roughly repeats the curve in Figure 8.6 (the collapse of an ordinary polymer chain). There is good experimental evidence that this stage ends up with a molten globule. It already has some vague features of the tertiary structure, correctly outlined but unfinished (e.g., particular positions of atomic blocks of the secondary structure, etc.) The second stage takes much longer. This is when all the structural details are properly set (e.g., positions of individual atomic groups, etc.)

However, the main question is still unanswered. How can all these processes create exactly the *right* structure in the end? Also, if we knew the primary structure,

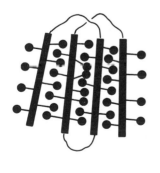

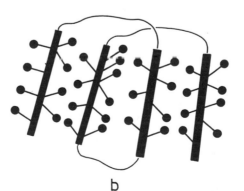

a   b

FIGURE 8.12
A sketch of the transition between (a) a native globule and (b) a molten globule.

could we in principle predict the tertiary one? The latter question is a practical one. If the answer were yes, it would be exciting news for genetic engineering. Moreover, we would no longer need complicated and expensive equipment for X-ray analysis of proteins. Unfortunately, this is far from being true. At present, we can guess the secondary structure with decent accuracy (i.e., α- and β-segments), but the tertiary structure remains a mystery. However, the study continues....

8.14 — *In Vivo, in Vitro, in Virtuo…*

Scientists are not unanimous even about how to begin to study life in general and protein folding in particular. Some say, "To learn about living things, we have to study them while they are still alive." This is called *in vivo*. Meanwhile, others argue, "Life is too complicated. We will understand nothing if we don't experiment with simple models." This is known as *in vitro*. The discussions have been going on for decades, causing mutual irritation, as well as progress on both sides. However, nowadays a third way has appeared, which could be called *in virtuo*. We are talking about the so-called virtual reality that does not exist anywhere but in the memory of a computer.

Let's give an example from a familiar field. Imagine a "polymer" that consists of 27 units (monomers). Suppose the monomers can be positioned at the vertices of a cubic lattice (Figure 8.13). In a close-packed state, such a polymer would occupy the volume of a $3 \times 3 \times 3$ cube (this is why we chose the number 27 in the first place). Now, it turns out that a "27-mer" can be arranged on a cubic

FIGURE 8.13
Heteropolymer of 27 different monomers can fill a $3 \times 3 \times 3$ domain on the cubic lattice.

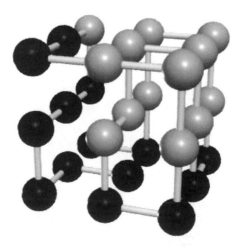

lattice in surprisingly many ways. There are about a hundred thousand possibilities (103,346, to be precise)! So you could say that there are 103,346 different globular conformations. Although the number is large, it is still possible for a computer to "churn out" the energies of the polymer in all its conformations.

How can this be done? Say there are monomers of q different sorts (we can think of them as different "colors"). How do they interact? Let's assume that, if two monomers of the same sort happen to be "neighbors" on the lattice, they attract each other with a certain energy, $-J$. Meanwhile, two neighboring monomers of different sorts repel each other with an energy $+J$. We could also look at various other kinds of interactions. As a result, we would find that, for very many primary sequences, one of the conformations has a much lower energy than any of the others. Could this be the actual reason the structure of proteins is unique?

Computer models (i.e., experiments *in virtuo*) can take us even farther, but not really that far, unfortunately. Suppose you wish to explore a longer polymer. The "magic" numbers that you have at your disposal are the following: 36, 48, 64, 80, 100, 125, ... A "36-mer" can be close-packed on a $3 \times 3 \times 4$ lattice, a "48-mer" can be fitted onto a $3 \times 4 \times 4$ lattice, and so on. No modern supercomputers have managed even to list all the possible states of a "64-mer." And you do need to list them all to discover the state with the lowest energy. (Nature may have learned somehow to disregard Levinthal's paradox, but we have not yet!) A "48-mer," for example, has as many as 134,131,827,475 close-packed states. This is too big a number to handle, in terms of calculating all the energies. Thus, a "36-mer," with its 84,731,192 close-packed states, and a "27-mer" are the only two models that are manageable so far (although there are some models on a flat surface; they are also useful and interesting).

Not only the difficulties but also the possibilities of the model grow really fast with the chain length. For example, "36-mer" is the shortest of the "magic" numbers that can have a knot, as shown in Plate 10. (By the way, are real proteins knotted? This is an interesting question, indeed. Apparently, some of them are.)

The models can be improved in a different way. Look at the picture Plate 11. Here is a lattice globule with a pocket, where we can put a "substrate." Plate 12 shows that this globule is able to "renature," that is, to self-assemble its correct structure. In this sense, it is indeed similar to a real protein molecular machine. Moreover, one can easily imagine the whole machine shop, as in Plate 13. In a way, this reminds us of a well-known joke about a theoretician who decided to do biology. This is how he started: "Suppose a horse has the shape of a cube, with a 1 m edge, and weighs 1 kg. . . ." Indeed, just compare Plate 5 and Plate 11! Well, whatever you say, together with the familiar *in vivo* and *in vitro* experiments, studies *in virtuo* are going on as well at full speed.

Look at the movie *"Folding"* on the CD ROM. In a way, it is similar to one of the other movies that perhaps you have already seen, the one called *"Collapse."* That was about a homopolymer, however. Now let's look at what happens when there are monomers of different types. Some of them attract more strongly, others more weakly, or maybe they even repel. Given that complex character of forces, the system appears to be deeply frustrated. By the way, this term, frustration, is not just our attempt to be simple, it is an exact scientific term. (Sometimes scientists pick good words to express their ideas!) So, monomers shown in the blue through violet color range repel and would prefer to move away from each other, but they cannot, since they are connected by the polymer chain; it is as if they are tied together by a rope. Other monomers shown in red through yellow attract and come as close as possible, but when they approach they pull, via the connections of the polymer chain, their repulsive counterparts. Thus, out of this complex game of pluses and minuses, of gains and losses, a very peculiar conformation can be born. Isn't this how real biological proteins work?

9

Dynamics of
Polymeric Fluids

9.1 — Viscosity

What do we mean by a polymeric fluid? It is a viscous liquid, made of heavily entangled polymer chains. In particular, it could be a polymer melt, a concentrated or a semidilute polymer solution. You can easily get a feel for what these are like. All you need to do is melt a piece of ordinary plastic, so that it starts flowing. Polymeric fluids are quite peculiar. In many ways, they are nothing like water or any other ordinary fluid that we are used to.

The first thing that strikes you is the high viscosity. It is normally much higher than for water. The physical cause of viscosity is internal friction. This acts between adjacent layers of the flowing fluid. Thus, we can say that internal friction in fluid polymers is greater than in water.

Let's bring some math into it. Figure 9.1 shows a very simple experiment. Some liquid is confined between two flat horizontal plates, a distance h apart. The lower plate is at rest, but the upper plate is moving at a constant speed v. How will the liquid at different heights move? A thin layer of liquid at the very bottom will stay at rest, due to internal friction. This layer is effectively "glued" to the lower plate. Similarly, the uppermost layer will be dragged by the upper plate

157

FIGURE 9.1
A liquid layer between two parallel plates. The upper plate is moving at a speed *v* causing a simple shear flow.

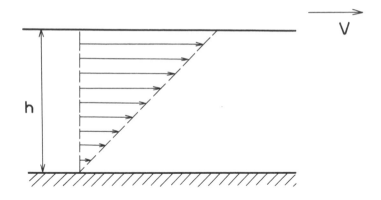

with speed v. The distribution of the speed across the liquid, shown in Figure 9.1, is called a simple shear flow. Of course, the upper plate does not move of its own accord, but is pulled by an external force. The question is: can we calculate this force f? Newton and then Stokes studied this problem in great detail. They came up with equation

$$f = \eta \frac{Av}{h}, \tag{9.1}$$

where A is the area of the plates and the coefficient η is the coefficient of viscosity, or just the viscosity. This coefficient gives a measure of how viscous the liquid is. For example, water has $\eta \approx 0.1 \text{ Ns/m}^2$ (at STP), whereas polymeric fluids may have η in the range 10^1–10^5 Ns/m^2 (depending on how long the chains are and whether there is any solvent).

9.2 — Viscoelasticity

The unusually high viscosity is not the only surprise that polymeric fluids offer. Another interesting and maybe even more important property is the viscoelasticity. Depending on the frequency of an external force, polymeric fluids can behave either like normal low molecular weight liquids or like elastic solids.

Here is a nice experiment to demonstrate viscoelasticity. Take a piece of the polymer called silicone. It is also known as "jumping putty." At room temperature it is in the viscous state. If you slightly tilt a jar of silicone, it will start flowing out, although rather slowly, as its viscosity is very high (Figure 9.2a). Now shape some silicone into a ball. It will bounce on the floor just like rubber (Figure 9.2b). (Remember the Native American balls made from natural, unvulcanized rubber?)

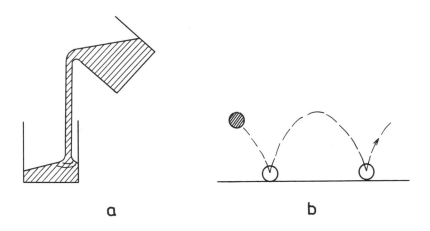

FIGURE 9.2
Experiments with silicone: (*a*) silicone flowing out of a jar; (*b*) a silicone ball bouncing on the floor.

a   b

How can we explain such a "double life"? The polymer behaves as a liquid when affected by gravity over a long period of time (Figure 9.2*a*). On the other hand, when the action of the force is very short (when hitting the floor, Figure 9.2*b*), the reaction is elastic. This is viscoelasticity. In general, viscoelastic bodies tend to show a viscous response to a slowly changing force and an elastic response to one which varies quickly.

There is another interesting experiment on viscoelasticity. Get hold of a cylindrical jar full of a concentrated polymer solution. Install another, smaller cylinder inside the jar, so that it can turn around the axis of the jar (Figure 9.3). Make the inner cylinder rotate at a constant angular speed ω for a while, and then suddenly let it go. Guess what will happen. Before stopping, the inner cylinder will turn back a little in the opposite direction! The angle α of this backward turn can reach a few degrees. Such unusual behavior is certainly a sign of viscoelasticity.

You could also look at a concentrated solution of polyoxyethylene or some other similar polymeric liquid (Figure 9.4). Tilt the container (A) and start pouring the polymer into another vessel (B) a little below it. Then, very gradually, making sure the polymer is still flowing, return the vessel A to the vertical position. It sounds amazing, but the polymer will not stop oozing from A to B, until A is completely empty! Thus, A and B behave as if they were connected by a tube (a siphon), but there isn't one—the role of the tube is played by the flow itself. Clearly, this would not work with water or another ordinary liquid whose behavior is mainly viscous. There must be some elasticity as well to achieve this effect.

All polymeric liquids are viscoelastic. This suggests that viscoelasticity is not caused by something special in the chemical structure; it is a universal property.

FIGURE 9.3
An experiment with a rotating cylinder demonstrating the viscoelasticity of a polymer solution.

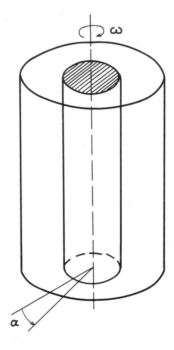

FIGURE 9.4
Two vessels, A and B, showing the siphon effect. The only connection between them is a jet of polymeric liquid.

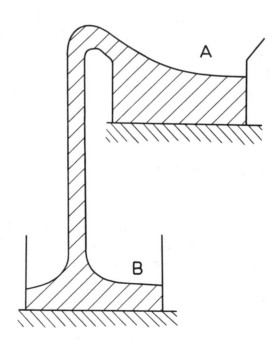

This is why theoretical physicists have flocked to study viscoelasticity and, in general, the dynamics of fluid polymers.

Is the viscoelasticity of polymeric fluids really so unexpected? Imagine a bunch of very long, very mixed up, and entangled chains. How does it flow? Obviously, a certain chain, if it wants to move, has to slither along a little wiggly corridor inside the bunch, undoing the knots on its way. This sort of picture inspired the theory of reptations (named after the snaky motion of reptiles). The first molecular theory of fluid polymer dynamics, it was developed in the 1970s by the physicists P. G. de Gennes from France, M. Doi from Japan, and S. F. Edwards from England.

Before we tell you more, we'll just make one comment. Some readers may have heard of L. D. Landau's scepticism about molecular theories of liquids. He thought it was impossible to create such a theory. All liquids are so different, they just don't seem to have enough in common for a common theory. Compare, for example, liquid helium and ordinary water. Another awkward thing about liquids is that there are no obvious small or large parameters. Most of the important dimensionless parameters are of order 1. This is a nuisance. As we have seen in Section 7.1, if you want to simplify a system's behavior and build an ideal model for it, you need some small or large parameters.

However, it is not so bad with polymeric liquids. There is a natural large parameter, the number N of monomer units in a chain. This is why it would be helpful to know how parameters such as the viscosity and molecular diffusion coefficients depend on N, in the limit of $N \gg 1$. The form of this dependence will determine how the polymer behaves. Such a situation allows us to use a very common approach of theoretical physics.

9.3 — The Reptation Model

Let's pick a test chain in a polymeric liquid. Imagine for a minute that all the other chains are "frozen" and cannot move. What can the test chain do in such a "frozen" jungle? It cannot go through other molecules. So it will be confined in a sort of tube formed by the neighboring chains (Figure 9.5). This is a fundamental concept. The chain cannot move through the walls of the tube, so all it can do is to crawl along. This is very clearly seen in a two-dimensional version in Figure 9.6. (In this figure, the "frozen" surroundings are modeled by fixed obstacles on the plane, which cannot be crossed by the chain while it is moving.) If there are no external forces, the motion along the tube is obviously purely diffusive. It is like

FIGURE 9.5
A polymer chain among
other chains in a
concentrated system.

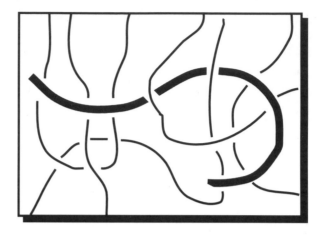

Brownian motion (see Chapter 5). The chain drifts in one or the other direction with equal probability.

Now, let us "defrost" the surrounding chains. Then more opportunities for the test chain will arise. Some of the neighboring chains will start moving away. Therefore, some constraints and entanglements that had formed the tube (Figures 9.5 and 9.6) will gradually disappear (or "decay"). However, as P. G. de Gennes showed, this effect is not important. The chain will snake out of the "frozen" tube much sooner than it takes the constraints to decay. This is why the motion in a fixed tube of obstacles is the main mechanism for the dynamics of a highly entangled chain.

The snakelike motion along the tube is called reptation, from the Latin *reptare*, "to crawl." The corresponding model of polymeric liquids is known as the reptation model.

FIGURE 9.6
A sketch of a polymer
chain in a network of
"frozen" obstacles (in
two dimensions).

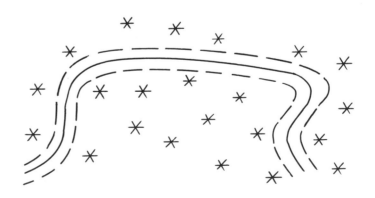

The concept of reptation is illustrated in the movie, "Reptation."

9.4 — The Longest Relaxation Time

It is interesting to see what we can learn from the reptation model. However, before we move on to this, we will look at a simple experiment. Take a polymer melt or a concentrated solution. At time $t = 0$, apply a constant elongating stress σ and measure the relative deformation $\Delta\ell/\ell$. If σ is small, the deformation will be proportional to the stress:

$$\frac{\Delta\ell}{\ell}(t) = \sigma J(t). \tag{9.2}$$

The function $J(t)$ is called the compliance of the polymer. On a logarithmic scale, it looks like the curve in Figure 9.7. After a sharp rise at the start, it reaches a plateau, $J(t) = J_0 = $ const. If we set this constant $J_0 = 1/E$, then, in the plateau region, we have

$$\sigma = \frac{E\Delta\ell}{\ell}, \tag{9.3}$$

which is Hooke's law (cf. (3.1)). Thus, the melt is elastic in this region, with Young's modulus E.

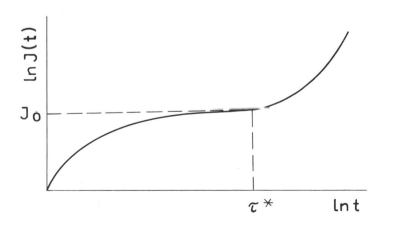

FIGURE 9.7
A typical J(t) for a polymer fluid.

Only after a long enough time, $t > \tau^*$ (Figure 9.7), does the deformation become irreversible and the polymer start to flow. In this case, the compliance is a linear function of time, $J(t) = J_1 t + J_2$, where J_1 and J_2 do not change with t. Let's compare this dependence with the definition (9.2) of the function $J(t)$. We can conclude that in the range $t > \tau^*$ the stress is no longer proportional to the strain, but rather to the rate at which the strain increases

$$\sigma = J_1^{-1} \frac{d(\Delta \ell / \ell)}{d\tau}. \tag{9.4}$$

This is the typical behavior of fluids. If you are not convinced, just compare equation (9.4) with the Newton–Stokes law (9.1). Of course, you need to bear in mind that equations (9.4) and (9.1) were written for different types of deformation. While (9.4) describes a one-dimensional elongation, (9.1) works for shear. So you need to match the two together. The tangential shear stress in Figure 9.1, obviously, corresponds to the ratio f/A. Similarly, the role of the relative elongation $\Delta \ell / \ell$ is played by the angle of shear γ (see Figure 9.1). It is defined so that $\tan \gamma = x/h$. (Here x is the displacement of the top plate from where it was at $t = 0$.) Hence, if γ is small, $\gamma \approx x/h$. Remembering that $v = dx/dt$, we shall get exactly the dependence (9.4) from (9.1). Another interesting result we can draw from this comparison is that the coefficient J_1^{-1} in (9.4) is the same as the viscosity of the polymer η. Thus,

$$\sigma = \eta \frac{d(\Delta \ell)/\ell}{dt}. \tag{9.5}$$

Let's summarize. At $t < \tau^*$ the polymer melt behaves as an elastic body (see (9.3)), whereas at $t > \tau^*$ it is rather like an ordinary fluid (see (9.5)). Just for comparison, we have also sketched the compliance function $J(t)$ for a typical nonpolymeric liquid (e.g., water) in Figure 9.8. You can see that this graph has no intermediate plateau region corresponding to elastic behavior (9.5).

The time τ^*, when the polymer's type of response to the stress changes, is called the longest relaxation time.

What came before the reptation model? Experiments like the one in Figure 9.7 used to be explained in the following way. A polymeric liquid was thought to contain some kind of effective cross-links. In contrast to the usual chemical cross-links (formed by chemical bonds), the effective ones do not live for long. They can only last for a period of the order of τ^*. Then they break (or "decay"), and new cross-links are created at other places, and so on. Thus, when $t \simeq \tau^*$, the cross-links do not have enough time to vanish. They hold the sample together,

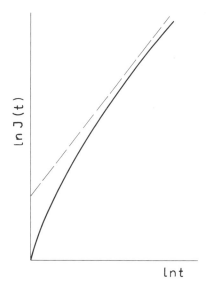

FIGURE 9.8
A typical dependence
$J(t)$ for a low molecular
weight liquid.

so it behaves like an elastic body. In contrast, when $t \gg \tau^*$, the cross-links start decaying, and the sample flows.

The reptation model makes this picture clearer. It tells us what these cross-links really are, at a molecular level. For example, pick two chains in a polymer melt. Both are both confined in their own fixed tubes. Suppose these tubes pass near to one another (Figure 9.9). Then the two molecules will have to enjoy each other's company, until one of them abandons the part of its tube that comes close to the other tube. You could say that, while the two tubes are next to each other, there is an effective cross-link in that area. However, as soon as one of the chains leaves the neighborhood area, the cross-link will disappear.

Now you can see the microscopic meaning of τ^*, which we introduced as a typical relaxation time of effective cross-links. The cross-links decay because of the chains' reptation, that is, because the chains slither out of their tubes. Therefore, τ^* gives an idea of the time it takes for the chain to abandon its original tube (where it was at $t = 0$). After this period of time, the chain finds itself in a brand new tube into which the random motion of its ends has moved it (Figure 9.10). You can say that the tube has been fully "renewed." All the original cross-links (i.e., the neighboring parts of different tubes) have totally vanished.

Let's go back to the simplest experiment shown in Figure 9.7 (where we applied a small constant stress σ at $t = 0$). What estimates can we make for the compliance in the ranges $t \gg \tau^*$ (viscous flow) and $t \ll \tau^*$ (elasticity)? From (9.5) and (9.2), we deduce that $J(t) \sim J_1 t \sim t/\eta$ for $t \gg \tau^*$. On the other hand, from (9.3) we have $J(t) \sim 1/E$ for $t \ll \tau^*$. Can we tell what happens for $t \sim \tau^*$?

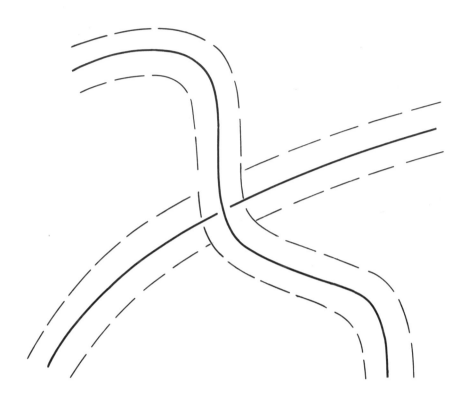

Obviously, the two estimates should merge smoothly into one another. This idea leads us to a very important relationship between the viscosity η, the longest relaxation time $\tau^{\star}$, and Young's modulus E for a network of effective cross-links:

$$\eta \sim E\tau^{\star}. \tag{9.6}$$

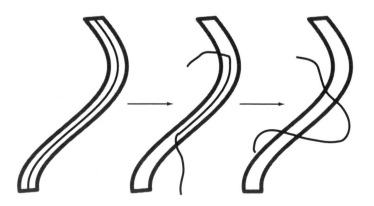

This relationship can help us learn about the viscosity of a polymer solution. In particular, we can use it to figure out how the viscosity depends on the number of monomer units N in a chain (in the limit $N \gg 1$). Presumably, we need to know first how E and τ^* depend on N. So we should venture a little investigation. Let's consider E and τ^* separately and concentrate on the case of a polymer melt to make it easier. In principle, the same sort of logic should apply to concentrated and semidilute solutions.

9.5 — Young's Modulus of a Network of Effective Cross-links

A network of effective cross-links behaves like a normal elastic network for $t \ll \tau^*$. We discussed the classical theory of high elasticity in Chapter 6. The Young's modulus of a network, as you remember, is of the order kT multiplied by the density of cross-links. (As usual, k is Boltzmann's constant, and T is the temperature.)

Thus, we have to work out roughly how many effective cross-links there are in a polymer melt. The tricky bit is to decide what exactly is an effective cross-link, and what is not. All the chains are highly entangled. An extreme view would be to regard any contact between a pair of chains as an effective cross-link. This is not completely illogical. Whenever a pair of chains come close to each other, their further motion is constrained (since they cannot go through each other). This is why the number of conformations allowed for each chain is much less than it would be in free space. You could model such topological constraints by effective cross-links.

What sort of picture would we really get if we replaced each contact between the chains by a cross-link? As you can imagine, it would be a very densely woven structure. All the cross-links would make it extremely stiff, so it would be nothing like an ordinary elastic body (i.e., nothing like our melt for $t \simeq \tau^*$). Suppose the chains in the melt are flexible, with Kuhn segment ℓ. The number of contacts per unit volume is approximately $1/\ell^3$. Let's accept for a moment the extreme view we have suggested. Then Young's modulus of the melt (i.e., of the network of effective cross-links) would be $E \sim kT/\ell^3$. How good is this estimate? We can check it if we use it to calculate the plateau value of the compliance, $J_0 = 1/E$, for various melts. It turns out that the answers it gives are far too high compared with experiments.

This is not surprising. In fact, there is a big difference between an effective cross-link and a mere contact between the chains. Look at Figure 9.11a, for example. The two chains pass near each other, so we can say that they are in

FIGURE 9.11
Contacts between
polymer chains:
(*a*) without a cross-link;
(*b*) with a cross-link.

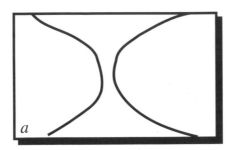
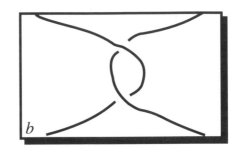

contact. However, this does not seriously restrict their freedom, that is, the choice of possible conformations. In contrast, the contact shown in Figure 9.11*b* is much more constraining. It is really just the same as a cross-link. In this case, the number of allowed conformations is obviously much reduced.

Thus, there are contacts and contacts. Not all of them play the role of effective cross-links. Taking this into account, let's modify the estimate for E:

$$E \sim \frac{kT}{\left(N_e \ell^3\right)}.$$

(9.7)

Here N_e is the average number of monomer units along the chain between two nearest effective cross-links. The parameter N_e is the only phenomenological one in the modern theory of polymeric liquids (i.e., it has to be found separately, from some other arguments or observations). Nobody has yet worked out how to calculate it from a knowledge of the microscopic structure. All we can say is that it must be somehow related to the ability of the chains to form knots with each other. Therefore, it must depend on the chain stiffness and geometry (e.g., whether it has any side branches, and so on). You can find N_e experimentally, from the value of Young's modulus corresponding to the plateau in Figure 9.7. Typically, N_e ranges from 50 to 500. In any case, $N_e \gg 1$. This confirms that only a small number of contacts work as effective cross-links.

How big is N_e in a highly entangled polymer? The reptation model talks about a chain in a tube. This only makes sense if there is a great number of effective cross-links per chain, that is, $N/N_e \gg 1$. We shall bear this in mind when deriving how the viscosity η and the maximum relaxation time $\tau^\star$ depend on N.

9.6 — The Tube

To find η and $\tau^\star$, we will pick a chain and explore its tube in more detail. The tube is created by other chains. If they come into contact with the test chain, they

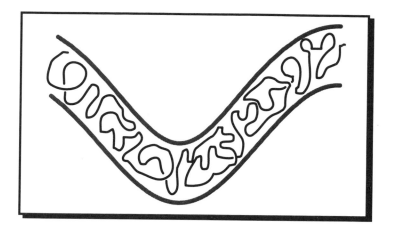

FIGURE 9.12
A chain in a tube.

act as obstacles to the chain's motion. However, we have seen that only a small proportion of such contacts can really limit the chain's choice of conformations. This proportion is of the order $1/N_e$. These are the contacts that can be regarded as effective cross-links.

Therefore, we come up with the following picture of a chain in a tube (Figure 9.12). First of all, there is the characteristic size $d \sim \ell N_e^{1/2}$, which roughly gives the distance between the two nearest cross-links along the chain. On the scale $r < d$, the chain does not "feel" the effective cross-links. So it has the full choice of allowed conformations. On the other hand, for distances $r > d$, the effective cross-links create the tube. This is why d must be the same as the diameter of the tube. Now we can regard the chain as a sequence of "blobs" of size d. Each blob contains N_e monomers and behaves as an ideal polymer coil.[1] They fill the tube, lining up along its axis. Hence, the total contour length of the tube axis is $\Lambda \sim (N/N_e)d$, since N/N_e is the number of blobs per chain. Remembering that $d \sim \ell N_e^{1/2}$, we get

$$\Lambda \sim \ell N N^{-1/2}. \tag{9.8}$$

Notice that this result for the length of the tube is much less than the full length of the chain $N\ell$. This is because $N_e \gg 1$.

---

[1] The blobs are ideal because the excluded volume interactions are completely screened in a polymer melt (see Chapter 7).

9.7 — The Dependence of the Longest Relaxation Time on the Chain Length

Now let's calculate the longest relaxation time τ^* for a polymer melt. As we have said, it is the time that a reptating chain takes to leave its original tube. To do this, the chain obviously has to diffuse along the tube axis by a distance of order Λ.

When a chain moves in a dense system (like a polymer melt), the frictional forces acting on each monomer are totally independent. Hence, the total frictional force experienced by the moving chain is simply the sum of the frictional forces on each individual monomer. How can we find these frictional forces on the monomers? Let's focus on one monomer; suppose it has a velocity $\vec{v}$. This is the velocity of diffusion, so it is not too high. (To be more precise, it is of the same order as the thermal velocity of the monomer.) This gives us the right to take the force $\vec{f}$ of viscous friction to be proportional to the velocity: $\vec{f} = -\mu\vec{v}$. Here μ is the coefficient of friction for a single monomer. Since the total friction is the sum over all the monomers, the same must be true for the coefficients of friction. The friction coefficients for the single monomers add up to give the friction coefficient for the whole chain. Say we have an N-unit chain crawling through a tube. Then its coefficient of friction μ_t will be just N times the monomer coefficient of friction μ: $\mu_t = N\mu$.

How do we normally describe diffusive (or Brownian) motion? An important quantity is the diffusion coefficient D. It determines the mean-square displacement $\langle x^2 \rangle$ of a Brownian particle over a period of time t (along one of the axes):

$$\langle x^2 \rangle = 2Dt. \tag{9.9}$$

(It is just because the motion is Brownian that $\langle x^2 \rangle$ is proportional to t; cf. (5.2).) How does friction come into this picture? Evidently, the greater the coefficient of friction μ for some particle, the lower will be the diffusion coefficient D, and vice versa. The exact relationship between the two was found at the beginning of this century by Albert Einstein and is called the Einstein relation. It states that

$$D = \frac{kT}{\mu}. \tag{9.10}$$

The physics of why the temperature T appears in equation (9.10) is quite clear. For given values of μ and t, the mean-square displacement (9.9) must be greater the higher the temperature, that is, the more intensive the thermal motion.

Now we can come back to the problem of the reptation of a long polymer chain in a melt. We will estimate the diffusion coefficient D_t, which describes lengthwise diffusion of the chain along the tube. According to (9.10),

$$D_t = \frac{kT}{\mu_t} = \frac{kT}{(N\mu)}.$$ (9.11)

As we know, the longest relaxation time $\tau^\star$ is roughly the time it takes the chain to diffuse along the tube by a distance equal to the length of the tube axis, Λ (equation (9.8); see also Section 9.4). Therefore, using (9.8), (9.9), and (9.11), we obtain

$$\tau^\star \sim \frac{\Lambda^2}{D_t} \sim N^3 \ell^{3/2} \frac{\mu}{(N_e kT)}.$$ (9.12)

We can tell from this that the longest relaxation time increases dramatically with N, the number of monomer units in the chain: $\tau^\star \sim N^3$. This explains why relaxation is so slow in polymeric liquids (compared to ordinary low molecular weight liquids). As a result, polymeric liquids have a long-lasting memory of the previous history of the flow. (If there were no such memory, for instance, the experiment shown in Figure 9.4 would be impossible.)

How much can the factor N^3 really slow things down? Let's make some estimates, to compare polymeric and low molecular weight liquids. We can rearrange equation (9.12) in the following way:

$$\tau^\star \sim \left(\frac{N^3}{N}\right) \tau_m,$$ (9.13)

where $\tau_m \sim \ell^2 \mu / kT$ is the microscopic relaxation time, typical of a low molecular weight liquid. Using (9.9), we can write $\tau_m \sim \ell^2 / D$, where D is the diffusion coefficient of a single molecule in such a liquid. Now we can see the meaning of τ_m. It is the time taken for a molecule to move a distance equal to its own size ℓ. Let's take some typical values $\ell = 5\,\text{Å} = 5 \cdot 10^{-10}$ m, and $D \approx 2 \cdot 10^{-7}$ m/s^2.

Then $\tau_m \sim 10^{-12}$ s. Thus, we have found the typical microscopic relaxation time for a low molecular weight liquid.[2]

According to (9.13), the longest relaxation time $\tau^\star$ of a polymer melt is a factor of N^3/N_e greater than τ_m. Suppose the polymer chains are rather long, $N \sim 10^4$. Then, using the crude estimate $N_e \sim 10^2$ (see Section 9.5), we obtain $N^3/N_e \sim 10^{10}$. This leads to a longest relaxation time of $\tau^\star \sim 10^{-2}$ s, which is a completely macroscopic value. It can be even bigger. Strong interactions between the molecules may sometimes increase the coefficient of friction μ. This will, in turn, increase $\tau_m \sim \ell^2\mu/(kT)$, and hence $\tau^\star$. The longest relaxation time may become as high as a few seconds or even more. This is just what you are likely to observe in experiments measuring macroscopic relaxation times of viscous polymeric liquids.

High values of $\tau^\star$ are responsible for viscoelasticity in polymers, which we can witness even in the simplest macroscopic experiments, like the ones described at the beginning of this chapter. If an external force is quite abrupt—that is, it acts for a period shorter than $\tau^\star$ (e.g., when a silicone ball hits the floor), there is no time for relaxation to occur. So the polymer behaves as an elastic body. On the other hand, if the force lasts for longer than $\tau^\star$ (like gravity making silicone flow out of a jar), the viscous friction comes into play.

9.8 — The Viscosity of a Polymer Melt and the Self-Diffusion Coefficient

Let's now use the reptation model to find the viscosity η of a polymer melt. We are going to use equation (9.6) and estimates (9.7) and (9.12) of Young's modulus E and the longest relaxation time $\tau^\star$. This gives

$$\eta \sim E\tau^\star \sim \left(\frac{\mu}{\ell}\right)\left(\frac{N^3}{N_e^2}\right). \tag{9.14}$$

---

[2] The estimate $\tau_m \sim 10^{-12}$ s gives a natural measure of time for liquids at room temperature. In fact, we could have obtained it in a different way. What we are interested in is the particle's displacement ℓ on the microscopic scale. To work it out, we can use either the diffusive relationship $\tau_m \sim \ell^2/D$ or the formula $\tau_m \sim \ell/v$, where v is the average thermal velocity of the molecules. Let's justify the latter. In dense systems, the size ℓ marks the border between two important length scales. On shorter scales, the motion of each molecule can be described accurately (very much like the free path of a particle in a low pressure gas). In contrast, at larger scales the molecules are engaged in diffusion. For light organic molecules at room temperature, $v \sim 500$ m/s. Thus, $\tau_m \sim \ell/v \sim 0.5 \cdot 10^{-9}/500$ s $\sim 10^{-12}$ s.

If the chains are long enough ($N \gg N_e$), the viscosity of the melt goes up quite rapidly as N increases: $\eta \sim N^3$ (just like the relaxation time).

We will also calculate the coefficient of translational diffusion D_s of the chain as a whole moving in the melt. While the chain completely leaves its original tube in the time $\tau^\star$, its center of mass must move by the distance $R \sim N^{1/2}\ell$, that is, about the size of a coil. The displacements of the chain during each interval of length $\tau^\star$ are statistically independent. This is why on a large time scale we can talk about the diffusion of the center of mass. It is just the same as the Brownian motion of a particle that has a mean free time $\tau^\star$. Between collisions it moves a distance of order R in a random direction (see Chapter 5). Therefore, according to (9.9) we can write

$$D_s \sim \frac{R^2}{\tau^\star} \sim \frac{N_e T}{(N^2 \mu)}. \qquad (9.15)$$

Thus, the reptation model predicts that D_s decreases as N^{-2} as the number N of monomers in the chain grows. When N is quite large, the diffusion coefficient is very low. As a result, if we bring two polymer melts together, they will tend to intermingle very slowly, even if the thermodynamics suggests that the mixed state is the most favorable one (i.e., if the two polymers are miscible).

9.9 — Experimental Tests of the Theory of Reptation

Do the main results of the theory of reptation (9.12), (9.14), and (9.15) agree with experiments? As for the estimate $D \sim N^{-2}$, the agreement is usually very good. However, it is not quite as pleasing with the power laws for the longest relaxation time $\tau^\star \sim N^3$ and the viscosity $\eta \sim N^3$. Most experiments indicate slightly sharper dependences: $\tau^\star \sim N^{3.4}$ and $\eta \sim N^{3.4}$. These are fairly close to the theory, yet not exactly the same. There have been many attempts to account for such a discrepancy. At present, the most widely accepted explanation is this. If the chains were infinitely long, experiments would give just what the theory predicts, $\tau^\star \sim N^3$ and $\eta \sim N^3$. But due to the finite length of the chains, we observe the index 3.4; the number of monomers N per chain is not big enough compared with N_e.

The reptation model is more powerful than you might think. You can get much more out of it than just the simplest basic laws for the viscosity, the longest relaxation time, and the diffusion coefficient of a polymer melt. The model also allows you to describe, for instance, the relaxation of a polymer after a stress has been released, or the response to a periodic force. As a result, you gain a fairly complete picture of the dynamics of polymeric liquids and of their viscoelasticity in particular.

The reptation model was the first to bring the large parameter N into play. As a result, a molecular theory of fluid polymer dynamics has been developed. All the previous theories of the dynamics of polymeric liquids were basically phenomenological.

9.10 — Reptation Theory and the Gel-Electrophoresis of DNA

Now let's dwell at some length on a rather unexpected, and probably one of the most important, application of reptation theory. Genetic engineering created an acute need for a high-precision technique for analyzing DNA. In particular, it was important to learn how to distinguish DNA strands that differed slightly in length, or in the amount of twisting or knotting (the latter two only make sense for ring-shaped DNA), and so on. The method of gel-electrophoresis has turned out to be amazingly suitable for that purpose.

The idea is the following. Suppose you have a solution containing the bunch of DNA molecules that you wish to separate. Spread this solution on the edge of a polymer network (i.e., a gel). In water, each of the DNA monomers will dissociate and acquire a negative electric charge. Therefore, if you place the sample in an electric field of the right polarity (i.e., between the plates of a capacitor), you can make the DNA chains move through the gel. Such motion is called electrophoretic. You only have to hope that the speed of the chains depends on their length and structure! If it does, the problem is solved. Different chains will travel different distances and will become separated (Figure 9.13).

Let us consider, for example, separating linear DNA chains of different lengths. Can we work out how the speed of the electrophoretic motion depends on the chain length, N? To get an idea, we will explore two limiting cases.

First of all, suppose the electric field is very strong. We follow the front end of a DNA molecule which moves forward and creates new bits of the tube. This end will spend more time traveling along the field direction rather than across it, or, even less likely, against it. As a result, the chain will tend to be stretched in the direction of the field (Figure 9.14). The stretching force f is proportional to N (because the total electric charge on the chain is proportional to N). Now, the coefficient of friction for the whole chain μ_t is also proportional to N, as we discussed when we talked about reptation. This is unfortunate. The speed of motion in a strong electric field is independent of N after all, $v = f/\mu_t$.

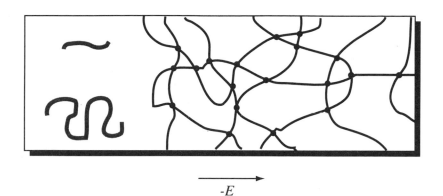

FIGURE 9.13
Explanation of how
gel-electrophoresis
works. DNA strands of
different sizes are
separated because they
differ in mobility.

$-E$

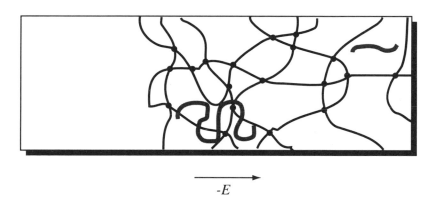

$-E$

Maybe we shall be more successful with weaker fields? But DNA molecules will not stretch in this case, but rather remain in the shape of Gaussian coils. The field pulls different parts of the chain in different directions (Figure 9.15.) We end up with a sort of tug-of-war game. Who will win? Obviously, it will be the end of the chain that happens to be farther forward in the field direction (and is therefore longer). The extra force it exerts is proportional to the displacement of this end, that is, to $N^{1/2}$. However, the coefficient of friction μ_t is still proportional to N. What a relief! The two dependences do not cancel out this time, and we find that the speed of the motion along the tube, v_t, is

$$v_t = \frac{f}{\mu_t} \sim \frac{N^{1/2}}{N} \sim N^{-1/2}. \tag{9.16}$$

(This is the speed of reptation. Don't confuse it with the speed of the chain as a whole, which we are still trying to find!)

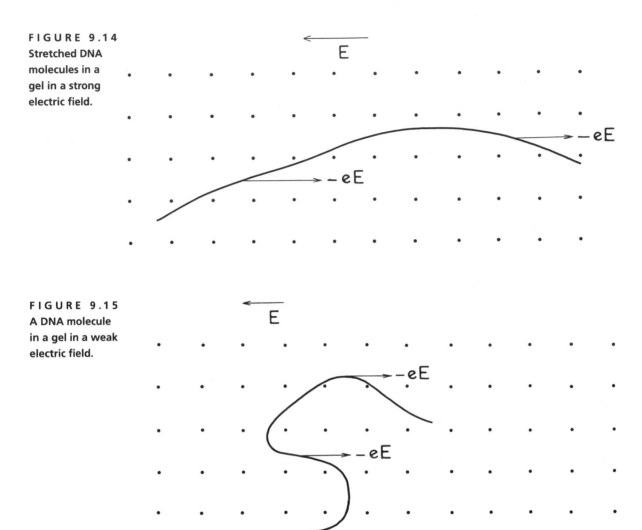

FIGURE 9.14
Stretched DNA
molecules in a
gel in a strong
electric field.

FIGURE 9.15
A DNA molecule
in a gel in a weak
electric field.

How fast will the molecule's center of mass move? Suppose the chain has crawled along the tube by a short distance Δ. We can represent this motion in a different way. Imagine that you chop a piece of length Δ at one end of the chain and stick it to the other end. The result will be the same. When you chop the piece, you transport it by a distance $\sim N^{1/2}$. The center of mass will move by a distance $\sim N^{1/2}\Delta/N \sim \Delta/N^{1/2}$. Hence, the speed of the center of mass is a factor of $N^{1/2}$ slower than the speed of reptation. Thus we find that the speed of the center of mass is $v \sim 1/N$ in a weak field.

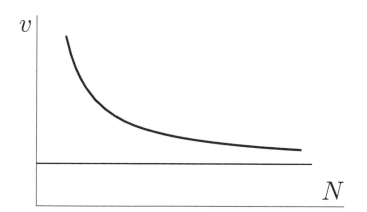

FIGURE 9.16
The dependence of
the speed of motion
v on the chain
length *N* during
gel-electrophoresis
of DNA.

A more accurate calculation confirms our answer. It gives the following formula:

$$\vec{v} = \frac{q}{3\eta} \left\{ \frac{1}{N} + \text{const} \left(\frac{Eq\ell}{kT} \right)^2 \right\} \vec{E}. \tag{9.17}$$

Here $\vec{E}$ is the electric field vector, q is the charge per unit length of the DNA chain, N is the length of the chain (measured in Kuhn segments), ℓ is the length of the Kuhn segment, η is the viscosity of the medium, and const is a number of order 1. The graph $v(N)$ is shown in Figure 9.16. When N is small, the dependence of v on N is rather strong. However, it flattens off as N increases, and becomes negligible. This means that only fairly short chains can be easily separated.[3] Of course, you can increase the threshold length if you reduce the electric field. However, this is not terribly helpful. In a very weak field, the whole process will be far too slow, which is inconvenient and may cause extra problems.

An interesting little trick has been devised to overcome these difficulties. The external field is periodically switched off (or rotated through 90°). The time taken for the field to go through one cycle should be roughly the same as the typical time

---

[3] Mathematically, v no longer depends on N when the second term in (9.17) becomes much greater than the first one. Then you can neglect the first term altogether. In practice, this is achieved at electric fields that are still not strong enough to stretch the DNA molecules fully. This is why our approach of guessing the dependence $v(N)$ from the two extremes is not very strict, yet the answer (9.17) is correct. You get the same in the proper theory.

of tube renewal, that is, $\tau^\star \sim N^3$ (see (9.12)). In this case electrophoretic motion will only occur for chains that have about the right N. Such a clever improvement has proved to work splendidly and to give extremely precise results.

9.11 — The Theory of Reptation and the Gel Effect During Polymerization

The theory of reptation helps us understand the gel effect during radical polymerization. We described how polymerization occurs in Section 2.7. Suppose we have added some initiator to the monomer, and the reaction has begun. At first, the growing chains appear in a kind of dilute solution, in which the monomer molecules play the part of a solvent. With time, more and more molecules of the monomer become involved in the reaction. The concentration of the chains grows, and they begin to overlap. This is when the solution becomes semidilute. From this moment on, the character of the chains' motion changes—they start moving by reptation. As we have already shown, this means that diffusion of polymer chains slows down substantially.

On the other hand, a polymer chain stops growing (i.e., breaks) when, as a result of diffusion, two free radicals at the ends of the chain happen to come together (see Section 2.7). Obviously, if diffusion slows down, such encounters of the chain ends become less frequent.

Thus, you may expect that as soon as the chains start overlapping, polymerization should proceed much faster. The chains themselves should be able to grow longer, because they do not break as often as before.

Indeed, all this can actually be observed, and is known as the gel effect during radical polymerization. The changes that occur are very dramatic. The rate of reaction jumps by a few orders of magnitude while the fraction of polymer increases only by a minute amount. This effect was noticed fairly long ago, well before the theory of reptation was proposed. However, it was the theory of reptation that enabled a proper mathematical description of the phenomenon.

10

The Mathematics of Complicated Polymer Structures: Fractals

10.1 — A Bit More About Math in Physics: How Does a Physicist Determine the Dimensionality of a Space?

A good starting point for another very interesting yet unfinished story is Brownian motion. As you remember, the displacement of a Brownian particle (or the end-to-end distance of a polymer chain) is proportional to the square root of the distance traveled (or the contour length of the chain). Surprising as it may seem in a book on polymers, the story is about the dimensionality of a space. Mathematicians have been studying this topic for nearly one hundred years and know quite a bit about it. However, it seemed of no importance for physics until very recently, after two books by the American physicist B. B. Mandelbrot appeared in 1977 and 1982 [13]. We shall avoid most of the math here and basically talk about the physics side.

The space we live in is, of course, three-dimensional. We know this because three coordinates, such as x, y, and z, are needed to describe any position. You might have also heard that time is often regarded as the fourth dimension. Thus, space-time is four-dimensional. A two-dimensional space is merely a plane, and a one-dimensional space is a straight line. However, it turns out that there are also objects with fractional dimensionality!

As proper physicists, we can imagine that an object consists of certain particles; we will call them simply "atoms." For example, we can picture a volume lattice (like a crystal one) consisting of atoms, or a flat film, or a straight-line chain. The dimensions of these objects will be 3, 2, and 1, respectively. We can make sure this is true in the following way. Take a sphere of the radius R, and count how many atoms of the object there are inside the sphere. Say this number is $N(R)$. For a volume lattice, $N(R)$ will be proportional to the volume of the sphere, $(4/3)\pi R^3$. Meanwhile, for a flat film, it will be proportional to the area of the crosssection through the center πR^2, and for a chain, to the length of the diameter $2R$. In all these examples, as you see, the dimension is given by some power of R. In the general case, we can write

$$N(R) = KR^d, \tag{10.1}$$

where K is a number independent of R. To get rid of this uninteresting constant, we take the logarithmic derivative of each side:

$$\frac{d \ln N(R)}{d \ln R} = d. \tag{10.2}$$

The quantity d defined by this formula is known as the dimensionality of the object. More precisely, it is the so-called fractal, scaling, or Hausdorff dimensionality. (In math, you may hear of others, such as metric, topological, and so on, but we won't talk about those.)

10.2 – Deterministic Fractals, or How to Draw Beautiful Patterns

"So what?" you may ask. "What's the use of equation (10.2)? Instead of the simple idea that there are length, width, and height in three-dimensional space, we now have a complicated formula, with a derivative and logarithms. What's the point?"

Look at Figure 10.1. These patterns are called Serpinski gaskets after the Polish mathematician who invented them at the beginning of 20th century. One can easily deduce the rule from the picture and use it to create all kinds of similar patterns. In all the examples, there are two kinds of bricks: grey ones and white

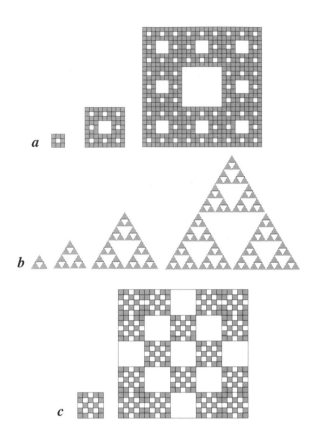

FIGURE 10.1
Serpinski gaskets—
simple geometrical
models of self-similar
fractal patterns.

ones. (If you use colored pencils or a computer with a color monitor, colored bricks can be used.) The elementary bricks can be squares (e.g., *a* and *c*), triangles (e.g., *b*), or any other sort of shape. Let's look at gasket *a*, for instance. It is quite easy to make the shape on the left from the white and grey squares. We can now think of this shape as a new large grey brick. Now let's make the same kind of shape using these large grey bricks, with white bricks of the same size. Obviously, we can carry on like this *ad infinitum*. As we make bigger shapes, not only are the white "holes" larger, but so are the grey areas.

As a matter of fact, people knew about this kind of patterns many years ago. Look at Plate 14. This figure shows floor mosaics of the church in the village of Anagni, Italy, which was built in the year 1104 and is made of Serpinski gaskets.

Now, suppose the grey bricks are "atoms," whereas the white ones are just cavities. Can we work out how many "atoms" there are in the system? Let's stick to gasket *a*: Having made ℓ steps, we shall have a square of side $3\,\ell$, therefore there are $(3\,\ell)^2 = 3^{2\ell} = 9^\ell$ original elementary blocks. There are 8 "atoms" (i.e., grey blocks) in the first figure. With every step, it gets multiplied by 8, so after ℓ steps it becomes 8^{ℓ}. Hence, if the size of a square is $R = 3^{\ell}$, then there are $N = 8^{\ell}$ "atoms" inside it. Simple algebra gives us $\ell = \log_3 R$, and $N = 8^{\log_3 R} = R^{\log_3 8}$. Thus, formulas (10.1) and (10.2) tell us that, in the case a, the dimensionality of the Serpinski gasket is $\log_3 8 = 3\log_3 2 \approx 1.89$. Similar calculations lead to the dimensionality $\log_2 3 \approx 1.58$ for gasket b, and $4\log_5 2 \approx 1.72$ for gasket c.

Thus Serpinski gaskets are a simple model of objects with a fractional dimensionality. Of course, the naive ideas of length, width, and height cannot possibly help us. They are just as absurd as the answers with a fractional number of people that some careless primary school pupils are known to come up with occasionally.

So what is the physical meaning of fractal dimensionality? Since $N(R) \sim R^d$, then the greater the value of d, the more "atoms" can be fitted into a fixed volume of the system, hence the fewer cavities there are. In this sense, you can say that the fractal dimensionality shows how "holey" the system is. More accurately, since the number of grey blocks in a Serpinski gasket on a plane is proportional to $N_{\max} \sim R^2$, what really characterizes the "holeyness" of the system is the value $2 - d$. Indeed, you can see by eye that the most "holey" gasket in Figure 10.1 is gasket b. And it really has the lowest dimensionality of the lot.

From what we have said, by the way, it follows that, if the system is situated on a plane, its dimensionality is $d \leq 2$. Similarly, in a three-dimensional space, $d \leq 3$, and so on. (If one fractal is placed on another one, $d_1 \leq d_2$!)

The question one may ask is this: Can the "holeyness" be described in a simpler way, for example, by means of density? Unfortunately, it can't. Take the gasket a. At step ℓ, the grey blocks are spread over a fraction $8^{\ell}/9^{\ell} = (2^3/3^2)^{\ell}$ of the area. This is what can be most naturally thought of as the density. As you can see, it tends to zero as ℓ grows. This is not surprising; it is a general law. The density is proportional to $N(R)/N_{\max}(R) \sim R^{d-2}$, that is, it depends on R and tends to zero as R increases, as long as $d < 2$.

It is very easy to write a computer program to draw Serpinski gaskets. All you really need to do is design a subroutine that composes the very first shape out of the elementary bricks. Then you can just keep calling this routine at each subsequent stage. In other words, you use the idea of *matryoshkas*, little traditionally Russian dolls that you put into one another. This principle works not only for Serpinski gaskets, but for many other patterns as well. Some of them are very beautiful, and they all are self-similar.

10.3 — Self-Similarity

Imagine an ideal geometrical straight line. Take a piece of it, say, 1 cm long. Now "zoom in" and look at this piece on a larger scale. What do you see? Again a piece

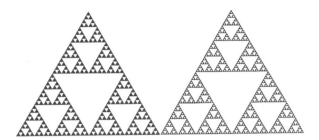

FIGURE 10.2
The idea of self-similarity. One of the two figures shown is the geometrically rescaled part of the other. It is really hard to say which is which!

of the same straight line. You can do the same thing with a geometrical plane. To see this another way, examine a real physical straight line drawn in pencil or made of string or wire. You can look at it at higher and higher magnifications, through a magnifying glass, then through a microscope, and so on. What you see remains a straight line, until you can start to make out its width or the "atoms" of which it is made.

Serpinski gaskets have the same property. There are two different gaskets in Figure 10.2. One of them was obtained from the other in two steps. First, the larger blocks were put together using the same "recursive" procedure as for the smaller ones, like in Figure 10.1. The resulting shape was then scaled down by a factor of 2. However, looking at Figure 10.2, you can hardly tell which figure is which! This is the property of self-similarity.

So we can see that the word *fractal* merely means a self-similar object.

Are there any fractals in nature? What is their significance in physics, if any? And what do polymers have to do with all this? Before attempting to answer, let's think about something slightly different—are there any geometrical shapes which are *not* self-similar? Of course, there are. Take, for instance, a circle. On a large scale it looks almost like a piece of a straight line, whereas on a small scale it is more like a single dot (Figure 10.3). Obviously, there is no similarity between the two whatsoever! This is precisely why the question about self-similarity is so important.

FIGURE 10.3
A circle at different magnifications. A circle is *not* self-similar.

10.4 — Natural Fractals

Plate 15 shows two photographs of what appears to be the head of a cauliflower. Actually, we took a picture of the whole head first, then cut a little floret out of the cauliflower and took a picture of that from much closer up. The screw in the photos shows the scale; if it were not there, it would be very hard to tell one object from the other. Moving the camera closer to the object is just the same as making a similarity transformation. Thus, a cauliflower is a self-similar object—a natural fractal.

This experiment had a nice side-effect—we had the tasty cauliflower left over! In our next experiment we shall end up with nothing but rubbish. Take a sheet of aluminum foil and cut it into little squares of different sizes. Then crumple them into little balls and measure the balls' diameters. (Now throw the balls away... or recycle them!) Figure 10.4 shows how a ball's diameter D depends on the size a of the square from which it was made. The graph is plotted on a logarithmic scale; that is, we have plotted $\ln D$ and $\ln a$ along the axes, rather than D and a. You can see that the experimental points fit nicely on a straight line. So, with a high degree of accuracy,

$$\ln D \approx \alpha \ln a + b, \quad \text{hence} \quad D \approx \text{const} \cdot a^{\alpha}. \tag{10.3}$$

The slope of the line is such that $\alpha \approx 0.88$.

What does it all mean? Let's go back to equation (10.1). The amount of aluminum (the mass or the total number of atoms) in a square of side a goes as a^2, that is $N \sim a^2$. Now, this aluminum is packed into a sphere with diameter D, so $R \sim D$. Therefore, since $D \sim a^{\alpha}$, we get $N \sim D^{2/\alpha} \sim R^{2/\alpha}$. This means that the fractal dimensionality of the crumpled foil is $d = 2/\alpha \approx 2.27 < 3$. The foil is not completely squashed; there are many little cavities left inside the aluminum balls. If we had a very sharp knife that could cut without squashing, we could chop the foil ball carefully and discover that the pattern at the crosssection is very much like the pattern of holes in a Serpinski gasket. Thus, crumpled foil is also a fractal.

Plate 16 shows a map of the Norwegian coastline between the towns of Bodö and Tromsö. We actually took three maps of this very place. One map was fairly detailed, with the scale 1 : 6,000,000. Meanwhile, two other maps were less detailed, 1 : 12,000,000 and 1 : 25,000,000, respectively. How do the three maps compare? The coarser maps leave out the finer details of the structure, such as little fjords and promontories. So the coastline drawn on a coarser map is not precisely the right shape; you could say the line is too thick to show all the little "wiggles" of the real structure (look at Plate 16 again).

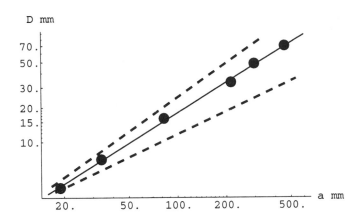

FIGURE 10.4
Log-log plot of the size
D of a crumpled piece
of aluminum foil versus
the piece's size *a*. Points
represent the data of
our measurements. The
solid line gives a best
fit of the form
ln *D* ≈ −1.18 + 0.88 ln *a*,
which indicates a
power law dependence
D ~ $a^{0.88}$, thus
manifesting the fractal
structure of crumpled
foil with fractal
dimensionality
d_f ≈ 2/0.88 ≈ 2.27 < 3.
Two dashed lines are
shown for comparison;
they correspond to
hypothetical systems
with fractal
dimensionalities 2
(unfolded foil) and 3
(dense piece of
material), respectively.

What can we tell about the length of the coastline from the maps? Of course, the answer depends strongly on how detailed the map is.[1] Measurements of the lengths of the coastline using the three maps mentioned above give 22, 10, and 3 cm, respectively. These results are shown as dots in Figure 10.5. The graph itself plots the logarithm of the line's length as a function of the logarithm of the map's scale. As you can see, the dependence is nearly linear:

$$\ln s \approx 1.4 \ln m + \text{const}, \quad \text{hence} \quad s \sim m^{1.4}. \tag{10.4}$$

From here we infer the fractal dimensionality d_f of the coastline to be approximately 1.4. What does it mean that $d_f > 1$? It indicates that the actual line is "thick" (see above). Why is $d_f < 2$? That is because the coastline is not really a surface (like a piece of fabric), but rather a border dividing the surface into two parts, sea and land.

So now, presumably, you will not find it hard to believe that there are also "rough" surfaces with dimensionality somewhat greater than two. And some cosmologists say (and they are not joking!) that the Universe has a sort of "foamy" form with dimensionality greater than four.

Now let us glance again at all the examples we have discussed. You may notice that they fall into two different groups. Some, just like Serpinski gaskets, fit perfectly on top of themselves when you subject them to a similarity transformation. This is just the way they have been constructed. Such fractals are known as deterministic and are studied mainly by mathematicians. In contrast, fractals from the

---

[1] In other words, the distance traveled between two points on the coast will be different for a ship traveling in the open sea and for a little boat or canoe that has to keep close to shore and follow all the shoreline's ins and outs.

FIGURE 10.5
The length of the coastline, measured from a map, depending on the scale of the map (on logarithmic axes).

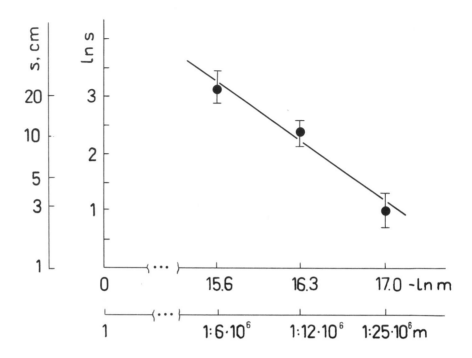

FIGURE 10.6
A Brownian trajectory is self-similar on average. The random walk (or freely jointed polymer) of 10^6 steps was generated computationally. In the main figure, every 10^3 steps are shown together as a single segment; there are $10^6/10^3 = 10^3$ of segments. In the inset, the "internal structure" of one segment is shown. It has 10^3 steps inside, and it looks very similar to the entire figure. *Source:* Courtesy of S. Buldyrev.

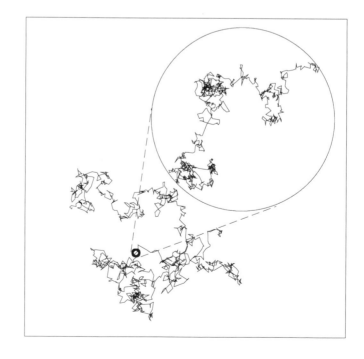

other group are self-similar only in some average statistical sense, judged by the general character of the pattern. The majority of "physical" fractals are like this.

Perhaps, for our purposes, the most important example of a fractal is the path of a Brownian particle which we discussed in detail in Chapter 5. How does an experimentalist detect a Brownian trajectory? He or she focuses a microscope, linked to a camera, on a fluid of suspended particles that is lit by a flash, say, every N seconds. The result on the film is actually not a trajectory, but merely a sequence of points. These points are connected by straight lines, and this gives a "coarse grain" model of the trajectory. This is the case shown in Figure 10.6. If we could make N flashes at 1-second intervals we would see what is shown in the enlarged portion of Figure 10.6. We would see a trajectory reduced in size, but generally of the same kind. Hence, a Brownian particle's path is also a fractal.

Brownian motion and geography, space physics and cauliflowers—it is really astonishing how far mathematics can generalize. However, let's get back to polymers. How can all this stuff possibly be related to them?

10.5 — Simple Polymer Fractals

The first "polymer" example is obvious. Indeed, we have already discussed how a polymer chain is bent and entangled, just like a Brownian particle's path. So it is bound to be a fractal!

So what is the fractal dimensionality of an ideal isolated polymer chain? The number of particles, that is, monomer units, is obviously proportional to the contour length of the chain, $N \sim L$. Hence, according to (4.3), the size of an N-monomer coil is $R(N) \sim N^{1/2}$. In other words, $N(R) \sim R^2$. Hence we obtain $d = 2$.

Thus, the fractal dimensionality of a free polymer chain turns out to be two. Although the chain is a sort of a line, its dimensionality suggests it must be more like a surface. To comprehend such a surprising result, imagine you flatten a polymer coil out on to a plane. This happens, for instance, when a polymer is adsorbed on to the surface of a solid. Alternatively, you could imagine random walks in two dimensions, for example, a rambler lost in a forest. If you have a long enough chain (or path), it will spread all over the surface more or less uniformly. (It is for precisely this reason that the rambler keeps coming back to places where he has already been!) You can think of this Brownian trajectory as being like a thread. It goes round and round, and gradually makes a piece of fabric, which is, of course, a two-dimensional object. In contrast, a molecule in the shape of an ordinary smooth line (such as a straight line) cannot possibly weave itself into any kind of fabric during adsorption. (By the way, this is exactly why it has only one dimension.)

We shall give some more examples below. However, even now we can conclude that the self-similarity and the fractal structure are not an exception but rather a rule in polymers and other complex systems.

We were talking in Chapter 7 about the swelling of a real (not ideal) polymer coil—due to the fact that every monomer is not an infinitesimal point, but a body of rather small, yet still finite, size. We have seen that the size of a swelling coil is $R \sim N^{3/5}$. A swollen coil is also a fractal, with a fractional dimensionality $d = 5/3$.

We have talked about linear polymers so far, but what can we say about branched ones? Of course, a lot depends on what sort of branching exists. For instance, if you have a long chain with tiny side bits, you can really treat it as a linear polymer, except with somewhat peculiar monomer units. A more interesting case, though, is a randomly branched polymer. You can imagine it in the following way. Suppose a polymer molecule is gradually growing. At each point it either stops or splits into two branches. Then you get a "tree," a bit like the one in Figure 10.7.[2] Even as early as 1949, B. Zimm and W. Stockmayer showed that the size of such a tree containing N monomers is proportional to $R \sim N^{1/4}$. Hence, a tree of this kind is a fractal, and its dimensionality is $d = 4$.

"This is rather extraordinary!" an observant reader might remark. "We have never come across a dimensionality greater than three before." Clearly, a straight line or a Serpinski gasket can be laid on a plane surface, since the dimensionality

FIGURE 10.7
Small piece of randomly branched tree. A much larger piece can be neither drawn on the paper nor fitted into space.

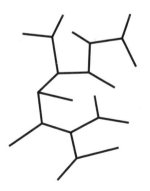

---

[2] There is quite a funny muddle of terms here: a *tree* is the generally accepted word for what we have described, yet its lattice model is known as a *lattice animal*. Does this indicate how well-informed physicists are about biology?

is less than two. In contrast, a three-dimensional object cannot be fitted on to a plane. In the same way, it is only natural to suppose that a four-dimensional tree would not really fit into a three-dimensional space. This conclusion is correct. In the process of branching polymerization, a tree becomes thick and either stops growing or stops splitting; in other words, it acquires longer and longer parts without any branches. (If you look around in a forest or a park, you can convince yourself that this is actually true for real trees.) For molecular trees, this thickening of the structure appears very important in some cases. For example, it causes blood to become denser and clot in the presence of air. Anyone who has ever been cut will appreciate this effect!

Of course, we did not need to talk about dimensionality to draw the right conclusion (about the thickening of a randomly growing chain in a three-dimensional space). Suppose a molecule consists of N monomers, and its size is about $\sim aN^{1/4}$. Each monomer occupies a small, yet finite volume v. Then the fraction of the volume taken up by all the monomers is proportional to $Nv/(aN^{1/4})^3 \sim (v/a^3)N^{1/4}$. So it increases with N indefinitely. On the other hand, in practice, a tree can only reach a volume fraction of order 1, and certainly no greater than 100%. So it is always true that $N < (a^3/v)^4$.

Which of the two types of argument is better, the dimensionality one or the density one? We won't try to conceal that there are currently different views about this. Even the two authors of this book have had a number of discussions on the question. You are welcome to eavesdrop on one of our conversations (the names of the authors are abbreviated to A. and S.).

10.6 — Why Worry About Fractals? (What the Two Authors Said to Each Other One Day)

A.: I wonder about this stuff on fractals. It feels like it's out in the cold. What really new ideas will readers have learned from this chapter?

S.: Why, they'll learn that such things as a Gaussian coil, a swollen coil, and a randomly branched polymer are all fractals. This is interesting in its own right. Mind you, there are more things in life than polymers. It might be interesting to hear about other fractals, the scale invariance of different objects, and the mathematical idea of fractional dimensionality.

A.: Perhaps you're right. But it doesn't follow from our text what you can do with fractals, what new problems they can help to solve.

S.: Yes, I see your point. But I don't think it's our fault! Suppose we weren't confined to simple examples, would we then be able to come up with such problems?

A.: The trouble is we wouldn't. As far as I know, no one has ever found anything new about polymers using fractal geometry. It was more about translating from an old language to a new one, rather than about deriving new things. Of course, it's a very beautiful language in its own right.

S.: Exactly! Remember Goethe's comment "Mathematicians are like the French..." that we put at the start of Chapter 5? But, to be serious, what I really want to emphasize is this. First, it is not just interesting—it is often useful to master different ways of describing the same thing. (Never mind that they are mathematically identical!) This is exactly what Richard Feynmann illustrated for the law of gravity, in his wonderful book, *Character of Physical Law* [14]. Second, there are loads of examples in physics where new achievements (and sometimes rather exciting ones!) were not expressed in the language of fractals, yet were very closely connected with them, due to the use of power laws and fractional dimensionalities.

A.: Yes, of course, it would be hard to disagree with you on that one. One cannot help referring to the work by the Nobel prize winner K. Wilson on the properties of strongly fluctuating systems. It turned out that the main problem was that the fluctuations do not "fit" into a three-dimensional space, but with increasing dimensionality, the situation simplifies. Even in just four dimensions it becomes trivial in some sense. So what did Wilson do? He looked at how the fluctuations behave for dimensionality $(4 - \varepsilon)$. If ε is small, than we are close to the simple case. He then found that the main features of what happens in reality ($\varepsilon = 1$) can be spotted even if you look at dimensionality 3.99.

S.: What a lovely example! I've just thought of yet another Nobel laureate, P.G. de Gennes. His ideas, all this stuff about scaling and blobs, that have proved so fruitful for polymers, are also connected with self-similarity and power laws, aren't they?

A.: Of course they are! But how much they have to do with all these light-hearted conversations about Serpinski gaskets and cauliflowers is a matter of opinion. Mind you, we haven't even explained how power laws come into the question.

S.: Good point. Well, let's give the readers an opportunity to judge for themselves what has to do with what, and to what extent.

10.7 — Why Is Self-Similarity Described by Power Laws, and What Use Can Be Made of This in Polymer Physics?

Starting with the very first equation (10.1), we have come across quite a few power laws in this chapter. Just look at Figures 10.4 and 10.5. The graphs are linear in logarithmic coordinates, which means that they represent power-law dependences. This leads us to the conclusion that the objects in question are fractals, that is, that they are self-similar.

As far as polymers go, the problem is to decide what monomer units make up the chain. "What a stupid question," a chemist would say. "If the chemical formula of a compound is A_N, and it is made up of lots of A molecules, then surely A is the monomer unit." However, why can you not connect the A molecules in twos, and then link all the $N/2$ dimers A_2 together? Or, alternatively, why not link together groups of three monomer units, etc., etc.? So

$$A_N = (A_2)_{N/2} = \cdots = (A_k)_{N/k}. \tag{10.5}$$

There is even no need to talk about synthesis in this case. Let's just take a chain, choose a particular piece of it, say consisting of g monomers initially, and regard it as the new "monomer unit" (see Figure 10.8). However, the properties should not depend on the particular way we describe the structure, that is, on g. How does this come about?

Let's calculate the distance between the two ends of a polymer chain. We already know that $R \sim aN^{1/2}$ for a Gaussian coil—see formulas (5.3) or (5.11). Now, if we are talking in terms of the "new" monomer units of size g, we should

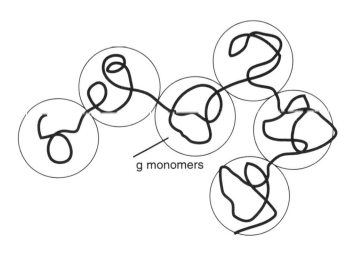

g monomers

FIGURE 10.8
An arbitrary number of links in polymer coil can be considered as a "new effective link."

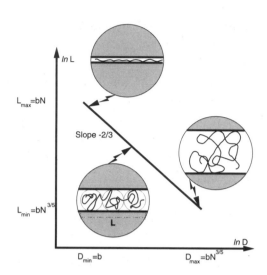

still get the same size R, which would be expressed by the formula of the same structure, but via the size of the new "monomer," a_g, and the number of new units, N_g:

$$R = a_g N_g^{1/2}. \tag{10.6}$$

So what are a_g and N_g? First of all, $N_g = N/g$. As far as a_g is concerned, we have to think in the following manner. Each new monomer is a tiny Gaussian coil in its own right. Therefore, a_g plays the same role for this tiny coil as R does for the normal one. Thus $a_g = ag^{1/2}$. Putting all these arguments together, we obtain

$$R = a_g N_g^{1/2} = ag^{1/2} \left(\frac{N}{g} \right)^{1/2} = aN^{1/2}. \tag{10.7}$$

Indeed, it does not depend on g!

We presume you might be interested in going through the same kind of proof for a swollen coil (which we discussed in Chapter 7; we proved then that $R = bN^{3/5}$), as well as for a tree ($R = bN^{1/4}$). You would then be able to see for yourself that the power laws do indeed correspond to self-similar objects, that is, to those which have, say, a g-unit organized in the same way as the whole thing (that is, it obeys the same power law as the whole chain).

Can we make any use of all this? Yes, of course, a lot! Look, for example, at the following problem. Take a real polymer coil. All we really need to know about it is its size, $R = bN^{3/5}$. Now try to "squeeze" it into a capillary of diameter D, as Figure 10.9 demonstrates. You may just look at Chapter 7, to see that even the simple question about a real polymer coil with excluded volume is very complicated. Moreover, if the coil is placed in a capillary, it is hard even to think how to start. However, imagine that it is de Gennes himself who is tackling it, using the idea of self-similarity. He is free to choose the monomer units in any way he likes. Suppose g is such that the size of the unit is equal to the size of the capillary, that is, $bg^{3/5} = D$, and therefore $g = (D/b)^{5/3}$. The technical term for such monomer units is "blobs." Such blobs, obviously, go one after another along the capillary, just like the cars of a train.[3] Let the size of each blob be D, and their number N/g. Then the polymer has a length

$$L \simeq D \left(\frac{N}{g} \right) \simeq ND^{-2/3}b^{5/3} \qquad (10.8)$$

along the tube. Is this all?? Yes, it is! Incredibly simple, and yet correct!

This points us to a useful habit that is certainly worth developing. As soon as you come across a formula, no matter how easy it is, it is a good idea to try and "get a feel for it" for some simpler limiting cases. What does this imply for equation (10.8)? There are two limiting cases here. Imagine a spacious tube, whose width is comparable to the size of the coil. Obviously, it should make no difference to the coil. This is exactly what we get from (10.8): If $D = bN^{3/5}$, then $R = bN^{3/5}$. (For even larger values of D, this equation would no longer hold, of course.) Now assume the opposite. The tube is as narrow as just a single monomer. Then the chain cannot possibly meander and has to stretch out into a straight line. This is also easily derived from (10.8). Indeed, if $D = b$, then $L = Nb$.

10.8 — Other Fractals in Polymers, and Polymers in Fractals

Perhaps our story about fractals is getting a little too long, but there are two more things that we must not leave out. Imagine little particles of soot flying out of a chimney, along with the smoke. These particles are sticky (if you don't believe

---

[3] A question that might crop up here is this. What if we look at the same capillary problem for an ideal Gaussian coil? In that case, just as usual, we would choose a blob of the size of the tube, that is, $ag^{1/2} = D$, so $g = (D/a)^2$. However, the blobs would then pass freely through each other. This is why equation (5.1) will not do for the random "walks" of the chain of the blobs in the tube, so we have to use (5.2) instead. This leads us to $L = D(N/g)^{1/2} = aN^{1/2}$. Thus, an ideal polymer squeezed into a tube does not become more stretched in the tube's direction.

it, touch them and then try to wash your hands!) When two bits of soot bang into each other, they get stuck together and make a larger particle. This flies on, picking up more and more sticky "mates" and growing in size. Thus, we end up with rather big flakes of soot. Some of them accumulate in the chimney, and others fly away. Yes, you are right if you are thinking that the structure of soot particles is self-similar; it is a fractal. Another example is snowflakes. They grow exactly in the same way from little "needles" of ice. (Talking about snowflakes, we are tempted to recommend to you some reading: *Letters to a Certain German Princess* by Kepler, who is indeed *the* Kepler, the discoverer of the laws for the orbits of planets.)

Many polymeric substances and materials are formed in a very similar way, when pieces of the substance stick together step by step. Take, for instance, blotting or filter paper, which consists of a complex fractal structure of pores and channels. In a way, it looks like a piece of crumpled foil, although the whole thing is smaller (and the fractal dimensionality is slightly different).

Perhaps we should say no more about such materials, but rather discuss how they are used in practice. That will lead us to the other side of the problem, "fractals and polymers."

A medium with complex branched pores can be used to purify and separate polymers. This is because chains of different lengths and different chemical structure move in different ways in such a medium (cf. gel electrophoresis, Section 8.10). Thus the question arises, how does an ordinary linear polymer behave in a fractal medium? Let's take a simple model: random walks across the little grey shapes on a Serpinski gasket. What will happen to equations like (5.2)? What will be the mean-square displacement, that is, the end-to-end distance, of the polymer chain? We hope that you are expecting to get a kind of power law for the dependence of the coil's size on the length of the chain, that is, $R \sim N^\nu$. The only question is, what is ν equal to? We suggest you play this game on a computer. It is quite fun as well as instructive. For our part, we shall only say that $\nu = d/2d_s$, where d_s is determined by what sort of sound waves would propagate along the fractal if it were made of little springs. It also depends on the heat capacity, C, of the "springy" fractal at low temperatures T ($C \sim T^{d_s}$). (It would be interesting to talk about all this at length. Unfortunately, our book is not elastic, even though it is polymeric!)

 In our movie *"Fractal Growth,"* we show just one example of fractal growth. It illustrates diffusion-limited aggregation. We did not include more movies about fractals here, because there are many of them elsewhere—in calendars, screen savers, and the like.

10.9 — Fractal Texts in DNA

We have been trying to keep our story up to date, at the very forefront of modern science. We thought it would be more fun to read, and it actually is much more fun to write! Thus in this section we will discuss a phenomenon discovered in 1992, with the main contribution from H. E. Stanley and his colleagues from Boston University. There are still fervent arguments about what they found. Some people have even suggested that there should be no effect at all. We think the effect exists, and for good reason. (We only mention the arguments to give you a flavor of how "hot" this developing field of science is.)

To begin, suppose we have a piece of text, say the text on this page of our book. In a way, it is just a sequence of letters. Now we would like to ask you a strange question: Is this sequence random? You may say: What rubbish! A random sequence of letters makes a random text! Do the authors really mean they have no idea what they are writing about??

We certainly hope we are making some sense in this book. But let's link some number, say $+1$, to all the consonants on this page, and -1 to all the vowels (plus punctuation marks and blank spaces). We get a sequence of numbers ξ_i. Now imagine a walker who makes one step to the right if $\xi_i = 1$, and one step to the left if $\xi_i = -1$. Can we picture this sort of motion? First of all, there are more consonants on the page (at least in the Russian original!), so the walker would tend to drift to the right (in the positive direction).

This is too obvious and not very exciting. However, it is posssible to get rid of the "drifting" effect. We can use equations (5.6–5.11) for our problem, which is actually simpler than the one we had before. Random walks are confined to a straight line and go only in two opposite directions, so we shall not even need any vectors. Let R_t be the displacement after t steps. Then $R_{t+1} = R_t + \xi_t$. In our previous problem we could square this equation straightaway because the average of $\vec{R}$ was zero. However, now we have a drift. (By the way, in the case of polymers, we would achieve the same effect if we pulled the chain by its ends in opposite directions.) So, bearing in mind the drift and taking the average, we obtain $\langle R_{t+1} \rangle = \langle R_t \rangle + \langle \xi_t \rangle$. From here $\langle R_t \rangle = t \langle \xi \rangle$ (obviously, the average value of $\langle \xi \rangle$ does not depend on t). Thus, the equation for the average displacement will be of the same sort as (5.1); it describes a simple uniform motion at a speed $\langle \xi \rangle$.

But what if we choose a different frame of reference, the one in which there is no such motion at all? Let's try it. Define the displacement as $S_t \equiv R_t - t \langle \xi \rangle$. Then $S_{t+1} = S_t + (\xi_t - \langle \xi \rangle)$. Obviously, S has no drift in this case. Its average value is zero. So we can turn back to the formulas from Chapter 5:

$$\langle S_{t+1}^2 \rangle - \langle S_t^2 \rangle = 2\langle S_t \eta_t \rangle + \langle \eta^2 \rangle, \tag{10.9}$$

where $\eta_t \equiv \xi_t - \langle \xi \rangle$. Now let's return to the main question: Is the sequence of letters in this text random? If the text is random, then no letter is determined by the preceding one. In other words, S_t and η_t are totally independent. Since $\langle \eta \rangle = 0$, we are left with the square root law, $S \sim t^{1/2}$, where $S = \sqrt{\langle S_t^2 \rangle}$. However, if the text is *not* random . . . .

It should be no surprise that all we have just said can be easily applied to DNA molecules. Indeed, DNA is a type of text written out in a four-letter alphabet. Suppose $\xi = +1$ for bases A and G, and $\xi = -1$ for bases T and C. Where does it lead us? The result is plotted in Plate 17 for the human beta globin gene. For comparison, we also show the what happens in the case of a totally random sequence of letters. The latter, of course, precisely follows the square root law. However, though the log-log graph for the gene is a straight line, the slope is far from 0.5. We find that $S \sim t^\alpha$, where $\alpha \approx 0.7$. This means that the DNA sequence is not random. Moreover, it is a fractal! (To see this, just look at Plate 17.)

At first, this sounds right. Indeed, it would be rather depressing to think that DNA is some totally random stuff, void of any sense. However, you may have second thoughts and wonder: Is that really all that DNA is about, just repeating the same motif over and over again, as fractals do? As a matter of fact, the greater part of DNA chain contains no codes for proteins. Moreover, no one even knows what that greater part is needed for (hence it is called "junk" DNA).

So what about texts that make sense (like DNA and, we hope, this book)? From the point of view of statistics, both DNA and normal languages look quite random; they have $\alpha = 0.5$. But meaning is too subtle a thing to be revealed at such a primitive level when we are only looking at alternating vowels and consonants. A mechanism for producing a meaningful piece of text has to be extremely complicated. The next letter of the text relies entirely on the actual context, not the physical side of the process, that is, what the letters and symbols are made from (be it ink on paper, nucleotides, or anything else). Just imagine a self-similar book: The whole book would look exactly like each of its chapters (which means that all the chapters would be identical!), and every chapter would look the same as any of its sections, and so on. This would not be very meaningful, would it?

What would be a good example of something purely meaningless? The most usual one is a random sequence of letters. As A. Eddington once suggested, such a sequence could be produced by a monkey if it were given a typewriter. (We already mentioned this when discussing random copolymerization in Section 2.5.) This poor monkey is mentioned in virtually every book on random processes, and yet nobody has ever given it the typing opportunity: the piece of literature the monkey would create might well have a fractal structure, even though it would almost certainly lack meaning. But the difference between being meaningless and meaningful is generally a very delicate matter and it has in fact to do with what we shall talk about in Chapter 11—so this is a good place to leave fractals.

11

Polymers and the
Origin of Life

11.1 — Why Do We Write About the Origin of Life in a Book on Polymers?

Every kitten has a cat mother and a cat father. Every human child has a human mother and a human father. Moreover, each of the child's parents also has (or had) parents, and so on.... Most people have probably been struck at some time by the frightening infinity of the past, which suddenly opens up so rapidly from such an ostensibly innocent thought.

There are many different theories about this tantalizing infinity. We shall not go deep into philosophy, of course.[1] However, regardless of our particular philosophical views, it would be quite interesting to know what existed "before,"

---

[1] The epigraph that Darwin himself selected for his book reveals his own views. It consists of three lengthy quotations, but, briefly, his ideas are the following. Divine Providence has no need to intervene in every petty happening. Instead, it governs the world by setting the general laws of nature. Thus, the theory of evolution has finally divorced science from philosophy. Now the two can more or less happily coexist and function in their own disparate ways, not hindering each other, and touching upon different facets of our amazing, inexhaustible world.

when there were no living organisms. Is it conceivable that something vaguely reminiscent of a simple biological molecule or structure could have appeared of its own accord in inanimate matter?

We are living at a very interesting time. At last, all such questions have ceased to give rise merely to groundless (and, therefore, rather impassioned) discussions among amateurs. Instead, they have become the subject of serious scientific studies by professional scientists. However, the most important thing for us, as far as this book is concerned, is that the physics of macromolecules has turned out to be at the very center of these studies.

It is very hard to write about the problem of the origin of life, especially when addressing a wide audience. It feels at times that it is more the prerogative of science fiction. There has been a fair bit of bluff and speculation on the subject, since so little is known for sure, and so much is still hidden in fog. On the other hand, as the Russian biophysicist L. A. Blumenfeld once said, "Living matter is the most interesting object for investigation by the living matter which is capable of investigating!" Encouraged by this, we cannot help devoting some space to the theory of evolution in this book on polymers. We must only promise to curb our excitement and to make an honest distinction between well-established facts and daring guesses: "Honesty is the first chapter in the book of wisdom" (Thomas Jefferson).

11.2 — The Picture of Biological Evolution from a Molecular Point of View

How does evolution look at the level of single molecules? Let's see what molecular biology has to add to the theory of evolution. Now that we understand the structure of molecules, we can talk about evolution in the language of math. Of course, the picture is far from complete; there is no mathematical description of the causes or motive forces of evolution. However, we no longer have to compare living organisms as such. We just examine what is written in their "identity cards," that is, in the primary structure of the biopolymers they contain. The wording in such "identity cards" tends to change from time to time; and this is exactly what evolution is at the molecular level.

Take, for example, the sequence of amino acids from the hemoglobin[2] of a human, a horse, and a shark. Out of 141 "letters," the human and the horse have

---

[2] Hemoglobin is a globular protein present in red blood cells. It carries oxygen from the lungs or gills to other organs and tissues.

only 18 that differ. Meanwhile, the human and the shark differ in about 79 of the amino acids. Presumably, this suggests that we are much more closely related to horses than to sharks!

If we checked other proteins, we would find that some types of proteins have more common features, and some have fewer, when compared between different species. For instance, the so-called histones differ far less than, say, fibrino peptides. This tells us something about evolution: The job that histones do (i.e., packing DNA chains into chromosomes) must have emerged much earlier than the responsibilities of fibrino peptides (blood clotting). By now, histones have been optimized (or, rather, reached the end of the road!). In contrast, fibrino peptides still have a lot of room for improvement.

Without doubt it is much easier, for scientists as well as for computers, to compare "texts" that are written out in a 20-letter alphabet than to deal with real creatures, either living or extinct. This is why primary structures are being gathered and cataloged with much vigor, and used for the reconstruction of "genealogical" trees.

Many interesting observations have been amassed. For example, imagine that you have a globular protein and replace one of the hydrophobic amino acids by another hydrophobic one (see Section 4.7). Or, you could just as well replace a hydrophilic amino acid by a different hydrophilic one. Suppose it is not an active center. Then, most likely, such a substitution will not cause much trouble. The three-dimensional structure of the globule will not be violated, and the protein will still be able to do its job. However, there is an unfortunate exception to this rule. Sometimes the substitution of one amino acid by another can be inherited (i.e., engraved in the code) and may cause a serious disease. Nevertheless, in general, the genetic code is amazingly resistant to misprints or interference. One particular property, noticed by M. V. Volkenstein, can hardly be accidental. If a DNA nucleotide is accidentally replaced by another, there is a more than even chance (in fact, about 2/3) that this will lead to hydrophobic-hydrophobic or hydrophilic-hydrophilic substitutions in the coded protein.

Can statistics be of any use to analyze the "letter" sequences in protein "texts"? A similar problem exists for human languages (if they are unknown, or you want to decode a message, etc.) The Russian language, for instance, was studied from this point of view by the famous mathematician A. A. Markov around the beginning of the 20th century. There are quite a few useful things that mere statistics can reveal, with no knowledge of the language whatsoever. (For example, you can distinguish poetry from prose.)

So what does "protein linguistics" tell us? It turns out that the protein "texts" are probably nearly random. Of course, an active center of a globular protein is formed by a few amino acids that are very important. You cannot replace them without damage. However, the rest (i.e., many tens of others) seem to be arranged fairly randomly along the chain—at least no special order has yet been found. O. B. Ptitsyn summed it up rather nicely: A protein is an edited statistical copolymer. What he meant was that proteins must have originally appeared with random primary structures. Over the long course of evolution, these structures have been only slightly edited, or corrected.

The idea that the primary structure of proteins should be nearly random was suggested as early as the 1950s, by G. Gamov. It was also at the very heart of the famous book by J. Monod, *Chance and Necessity* [10]. It has turned out to be much harder to find signs of regularity in proteins, that is, to discern the fingerprints of editing.

Having said that, there are new, much more sophisticated methods of tackling protein linguistics. They appeared as recently as in 1994 and have revealed some very subtle deviations from randomness. The current explanation is this: These deviations are the "fingerprints" of a certain prebiological process that led to the formation of proteins.

Another riddle is the "fractal linguistics" of DNA, that is, the meaningless part of DNA texts that we mentioned at the end of Chapter 10. (See Section 10.9.) Where does it come from? At present, nobody can tell for sure. Some people think that the fractal bits reflect the history of evolution. If DNA structures developed gradually, step by step, it might have been a process similar to what a modern computer does when building up fractals (i.e., little shapes are grouped into blocks, which, in their turn, are used to make even bigger blocks according to the same rule, etc.). Another opinion is that the "meaningless" part of DNA is needed to retain the structure of the chromosome (which is a DNA globule). It may be just like in proteins, where a sequence of monomer units determines the tertiary structure of the protein. If a chromosome has a fractal three-dimensional structure, then to retain it a fractal sequence is needed. However, all these ideas are no more than hypotheses.

Thus, at the molecular level, biological evolution is reduced to changes in the primary structure of biopolymers.

An important thing for physics is that a DNA double helix's shape, and hence its energy, do not really depend on the sequence of monomers. (This is because the monomers are mutually complementary and are hidden inside the double helix.) It is for precisely this reason that the DNA "texts" can be altered. Otherwise, the result of evolution would be not the best-suited organisms, but merely DNA molecules with lower energy. In contrast, the secondary and tertiary structure of proteins strongly depend on the primary structure. This is what allows different proteins to carry out so many different functions.

One last comment. Don't think that evolutionary changes in biopolymer "texts" are just restricted to simple point mutations, that is, to replacements of single "letters." If they were, evolution would proceed unbearably slowly, or, more likely, would not happen at all. In fact, there are special mechanisms to modify whole "words" or "phrases" in these texts. Would you like to know how they work? For an introduction, we recommend reading reference [9].

Lying in wait for us is the next question: How did it all start? We have talked a little about the molecular picture of the evolution of life, in which less complex forms develop gradually into more complex ones. But where did the simplest forms of life spring from? In the spirit of Darwinism, the answer is obvious: They evolved, step by step, from the inanimate world. But how did this happen? There are no surviving witnesses, of course (although there are scraps of apparent evidence)! Therefore we cannot rely on observations, but have to turn to theory and experiments. So what do we know about evolution before life came on the scene?

11.3 — The Conception of Life and the Evolution of the Universe

Some scientists have suggested that life did not occur spontaneously on Earth, but was carried here from outer space. However, even if there were some firm evidence for this idea (which there probably isn't), it would not help. It just moves the goal posts. We would still have to answer the question of how life appeared out there. Since we have started talking about the cosmos, let us recall that the estimated age of the Earth ($4.5 \cdot 10^9$ years) is comparable to the estimated age of the whole Universe (10^{10} years). This is why it is natural to regard the appearance and evolution of life on the Earth as a part of the Universe's evolution.

The current, firmly established view is that the Universe started about 10 billion years ago in a "Big Bang." At the very beginning, the Universe was very small, unimaginably dense and hot, and all was light (photons). The pressure of light caused the Universe to expand. While expanding, the Universe became more rarefied and cooled down (and this seems to be continuing). As it cooled, other particles started to materialize, first electrons and positrons; then protons, neutrons, and their corresponding antiparticles; later atomic nuclei; and so on. If you are interested in how scientists have found all this out, we recommend the book by S. Weinberg entitled *The First Three Minutes* [15]. We cannot do justice to this topic here, although the basic principle is very simple. Each particle "con-

FIGURE 11.1
A demonstration
showing that the
dimensionality of a
space can gradually
change as you move
through it.

denses out" at the maximum temperature it can stand without "falling apart."[3] Note especially that at each temperature all the energetically possible particles are formed.

The Universe is not uniform. Even now it contains galaxies and intergalactic space, stars and interstellar gas, and so on. Even the dimension of space is probably different in different parts of the Universe! (You can visualize this with the following demonstration. Make a tube out of paper and stick it together as shown in Figure 11.1, so that the diameter varies from 10 cm down to 1 mm. Creatures 1 cm in size inhabiting this strange universe will think that it is one-dimensional at one of its ends and two-dimensional at the other.) You may ask: Why do we live under precisely these conditions and in exactly this part of the Universe? The answer is given by the so-called anthropomorphic principle, which is well accepted by cosmologists. It says that we only witness particular types of processes, since all the other types can have no witnesses (as expressed by A. L. Zel'manov). In other words, life could have only emerged and evolved in places where there were suitable conditions for it.

11.4 — Chemical Evolution on the Early Earth

What is so special about our part of the Universe? Once upon a time, stars appeared near here, and their innards started to produce various kinds of atomic nuclei. Then the gas around the stars cooled down to such an extent that the planets could condense out. There various molecules were formed, and chemical reactions began. This is how chemical evolution arose. On our planet it is thought to have started 4.5 billion years ago.

---

[3] Compare this with some facts from everyday physics. Take water. At 287 K (0°C), ice crystals "fall apart" (melt); at 387 K (100°C), water droplets "fall apart" (evaporate); at 10^4 K molecules "fall apart" into separate atoms; at 10^5 K atoms lose their electrons, "falling apart" into plasma, etc. The higher the temperature, the smaller the units that can exist.

In a sense, you can say that the Earth with a surface temperature $\sim 10^2$ K reminds us of a water wheel. What makes the wheel turn? Obviously, it turns because it is in the way of a stream of falling water. In the same way, the Earth is in the way of the "stream" of light that rushes from the hot Sun (with the temperature $\sim 10^5$ K) through cold outer space (where the background radiation has a temperature of about 3 K). We all know that this "stream" is still turning the "wheels" of life. But how did it start working on the primordial Earth?

What was there on the Earth at that time? There was the atmosphere, water, and land. And there was certainly light coming from the Sun. Violent processes were occurring: Winds blew, waves battered, rivers rushed, thunder and lightning rent the air, and volcanoes exploded. The atmosphere consisted chiefly of the simplest gases: nitrogen, carbon monoxide, steam, and hydrogen. (The latter rapidly escaped from the outer layers of the atmosphere.)

One might wonder what sort of miracle could have happened in such severe conditions to bring about life. Nitrogen N_2 and carbon monoxide CO, together with hydrogen H_2, gave birth to ammonia NH_3 and methane CH_4 (with the release of water). Other gases.... well, do we need more examples? After all, it is not that unlikely that any fairly simple compound (not a polymer, of course!) could have eventually been created, even given the scanty choice of original elements. There were just so many possibilities how it could have happened. In one or other place, in deep or shallow water, in the air, or in the sand. At any time, during millions of millennia ($\sim 10^9$ years). In larger or smaller quantities. Through one or other chain of chemical reactions. If a particular reaction needed energy, there was no problem with that. It could have been supplied by the ultraviolet irradiation from the Sun, electrical lightning discharges, hot volcanic products, shock waves, and so on. Clays could have played the role of enzymes. There were 24-hour cycles of different conditions of light, temperature, and humidity.

No doubt, this offers a vast field for fantasies, but there are just as many opportunities for scientific research. There are lots of questions to be answered. In what quantities were particular compounds produced, how fast, what mixed with what, and which substances were separated? To get some idea of all this, special laboratory experiments have been set up. A hermetically sealed vessel is filled with the right mixture of solid, liquid, and gaseous ingredients. Appropriate light and electrical discharges are provided. All the conditions (such as the brightness of the light, the average frequency of "lightning" discharges, etc.) are chosen in such a way that, say, a week of the experiment would be equivalent to some 50,000 years in reality. At the end, the resulting mixture is analyzed carefully. What do they find? The answer is: nearly everything. You can learn all your chemistry if you just go through the list of the final products. Even such complex yet irreplaceable "building bricks" of nature as amino acids are obtained in fairly respectable quantities.

To summarize, the chemical evolution of the early Earth obeyed the same principle as the cosmological evolution mentioned before. At each stage, you get the particles (or molecules, as long as we are talking about chemical evolution) of all the possible sorts allowed by the energy conditions. In the experiments by Fox (1968), even protein-like polymers, the so-called proteinoids, were obtained.

Does this mean that, among all the other things, the proper proteins and DNA could have appeared spontaneously? And so there is nothing more to say about the origin of life but that it was just a pure accident, that's all? Of course not. What we have been talking about so far only holds for nonpolymeric molecules.

People with very keen insight have never really doubted that biological evolution was far more complicated than chemical evolution. As an example, let's just quote the great German scientist and philosopher Immanuel Kant (1724–1804). (By the way, he was the first to suggest a scientific theory of the evolution of the Universe, which was identified with the Solar System at that time. In particular, he proposed that the planets were formed by the condensation of hot nebulous matter.) This is what Kant wrote about evolution, more than 200 years ago: "It is easier to understand the creation of all the celestial bodies and the cause of their motion, in other words the origin of the whole present-day organisation of the universe, than to find out by means of mechanics how a little blade of grass or a caterpillar appeared." Today we know that what could have really changed the whole character of evolution was the process of polymerization.

11.5 – Prebiological Evolution: Polymers "Scoff" Each Other's Food

The mixture of early evolutionary products on the Earth is usually called the primordial soup. In this soup, there were monomers that could join up with each other, given favorable conditions. They formed polymer chains. This is a well-established fact, proved by laboratory experiments. Moreover, some of the polymers formed had a slight ability to act as enzymes (see Section 4.7), which is a most important feature of modern biopolymers. This has also been confirmed by experiments. However, what happened next is much less clear.

There is a riddle you often hear: What came first, the chicken or the egg? You may think it is a joke, but it is really another wording for the problem of the origin of life. Molecular biology has not solved it yet, but has rephrased it once more: What was first, a DNA molecule carrying a blueprint of how to make a protein, or a protein that synthesized a DNA molecule? This question reminds us of the most important function of modern biopolymer systems, namely, cyclic reproduction.

Let's consider a very simple model that might help us out of the maze. Suppose a polymer chain is created accidentally in the primordial soup. Entirely by chance, it is a weak enzyme and can speed up the production of copies of itself. In other words, it stimulates the monomers from the soup to join each other and to form chains of the same primary structure. To be frank, this picture takes us away from the island of well-established facts into the ocean (or soup?) of rickety conjectures. The main thing is to remain sober, and discuss it as a possibility, rather than as reliable truth. This is what theorists often have to do anyway, whenever they face a stubborn problem.

So what would happen if a hypothetical primary structure could make copies of itself? As soon as such a structure appears in the soup, it will cause a kind of explosion! The chain-enzyme will start making its own copies, one after another. All the newly born identical chains will immediately get involved in the same job. It will be a snowballing process, following a geometric progression, until all the monomers in the soup are used up.

To get closer to what really happens in the living cell, let's make our model a little more general. Suppose the molecules of a certain type A (e. g., DNA with a particular primary structure) catalyze the synthesis of the molecules of type B (protein). Molecules B, in their turn, catalyze the synthesis of A. We could even think of a more complex system, when A makes B, B makes C, C makes..., ... makes A. The German biophysicist Manfred Eigen named such a system a hypercycle. Obviously, the reproduction of the molecules taking part in the hypercycle goes in the same way as the self-copying of the chain in our first example. As soon as you have the starting molecules, it runs out of control and only finishes when all the building material is used up.

Do we have to restrict ourselves to just one such hypercycle (or one such chain)? What if there were two or more? Then things will become much more interesting. If one of the hypercycles (or one of the chains) has a more efficient catalysis, it will reproduce itself more rapidly. Hence, there will be more chains of the corresponding type. Meanwhile, the stock of monomers in the primordial soup is limited and is shared between all. Moreover, the chains tend to break spontaneously from time to time. The monomers that are released when some chains are broken are then reused to build new chains (or to complete a hypercycle). However, it is most likely they will be used by the most efficient one. This is why we shall end up almost exclusively with the representatives of one sort.

This story is remarkably similar to how Darwin's "survival of the fittest" is often described, at a primitive level. The monomers play the role of food, and self-copying chains or hypercycles play the role of living beings who reproduce themselves given enough food. As a result of the competition for fodder, only the most gluttonous and prolific species will survive.

M. Eigen, whom we have already mentioned, his colleagues, students, and followers have investigated many models of the "evolution of polymers" [6]. They dressed these models up in a mathematical form. The story about reproduction and catalysis was turned into differential equations. All this makes beautiful science. It enables us to trace how the best molecules "scoff" their neighbors' food. (By the best, we mean, of course, the best for evolution, i.e., those that can survive because they have more efficient catalysis.)

Does this mean the problem of the origin of life has been solved? Polymer chains emerged spontaneously in the primeval soup, then a kind of "beauty competition" was held among them, the best ones were selected, and those lived happily ever after? No, things are much more complicated than that and, fortunately, more interesting.

11.6 — Primary Polymerization: Can *War and Peace* Be Written by Chance?

Out of m monomers of different sorts you can make m^N different polymers each of length N. They will only differ in the sequence of monomer units, that is, in the primary structure. However, they really will be distinct polymers, as we know, for example, that different proteins differ greatly from each other in the functions they carry out. Thus, the number of possible polymers grows exponentially with their length, $m^N = \exp(N \ln m)$.

Physicists know that if there is a large parameter in the argument of an exponential, it needs to be treated with care. We have already come across such a situation in this book. The expression m^N is not as simple as it looks. It tells us that the Earth is not old enough, nor is there enough material on it, for all the possible sequences of monomers to have been tried in selecting the best. Therefore the principle to which we referred when we talked about cosmological and chemical evolution no longer holds. (As you remember, it was that all possible sorts of particles are formed if there is enough energy for them.) This takes us to a new, utterly different stage of evolution. Our principle, having once been broken, does not work for any further stages either.

Let's make some estimates to confirm what we have just said. For example, how many protein chains of a given length (say, 200 monomers) can exist? For proteins, $m = 20$, and if we take $N = 200$, we shall have $m^N = 20^{200} =$

$10^{200\log 20} \approx 10^{260}$. This number is ludicrously huge. Even if the whole surface of the Earth (roughly $5 \cdot 10^8$ km^2) were covered with a 10-km thick layer of protein-like polymers, we would "only" manage to fit in about 10^{34} chains. (This is because the volume taken up by each 200-monomer chain is roughly 10^{-20} cm^3.) Now imagine that every molecular collision (lasting for about 10^{-11} s) throughout the history of the Earth ($4.5 \cdot 10^9$ years) led to renewal of the primary structures of all the chains (which sounds even more unlikely than anything we have said so far!) Even then "only" about 10^{28} possibilities would have been tried out by now. Hence, our generous overestimates give an answer in the order of $10^{34} \cdot 10^{28} = 10^{62}$, which is still far too far from 10^{260}. As you see, exponentials are not to be trifled with!

Thus, there were too many possibilities for the chains to try out. As a result, not all the polymers but only a tiniest fraction of them could take part in the "beauty competition" held by evolution. As we shall see later, this suggests that it was precisely at the stage of polymerization that the first "life-like" features started to show up. How could evolution have progressed when there was no more opportunity to make the best choices?

First of all, let's try to answer the following. How many primary structures, out of the total number m^N, are likely to have the special gift for self-catalysis (or mutual catalysis, in the case of hypercycles)? Is it many or just a few? One certainly needs to be careful with the words "many" and "few" in this context. A couple of thousand, million, or billion are nothing compared, for instance, with the specter of 10^{260}. If we do not want to reach an impasse in an attempt to trace our own pedigree, we have to assume that nearly every primary structure has this useful gift. The meaning of "nearly every" is certainly just as relative as that of "many." Maybe only one single chain out of a million, or a billion, is a catalyst, but we think that the total number of catalysts is comparable to 10^{260}. You can say that the very existence of life on Earth is itself a strong experimental evidence for such an assumption!

There is more specific evidence as well. First of all, we have already mentioned that "proteinoids" manifested a weak catalytic activity. They were obtained in Fox's experiments modeling chemical evolution. Second, primary structures of the modern proteins are still nearly random, even after $4.5 \cdot 10^9$ years of biological evolution. (Remember in Section 11.2 that we said that "a protein is an edited statistical copolymer.")

In addition to all this indirect evidence, it would be nice to find some more convincing proof. Can we calculate the probability that a sequence of monomers, picked at random, will be a functioning polymer? This is a similar task, in a sense, to the prediction of the tertiary structure of a globular protein (see Chapter 8). If you remember, we were interested in the possibility of deriving a three-dimensional structure from a primary one. Neither problem has been solved, but both are intensively studied. Several divergent approaches are being explored. Who knows, maybe some of the readers of this book will manage to clarify the matter!

Just in case anybody wants to give it a try, we ought to explain one more thing. This problem is related to what we said about the difference between meaningful and meaningless texts. As we have just seen, it is enormously unlikely that a meaningful text of any sensible length will appear entirely by chance. Now we are asking for even more! Not only do we expect the text to appear, but we are also awaiting a reader who will decide whether the text makes sense or not. Moreover, in this case the text and the reader are almost the same thing, in some sense....

Back to evolution. We have decided that not all the possible chains were synthesized. Far from it. Anyway, what sort of chains are we talking about? Over the last few years, more and more evidence has been gathered that RNA played the main role. RNA can carry out DNA's "instructive" function and could presumably work as a decent catalyst (although nowadays this is the job of proteins). Moreover, it is not quite clear why RNA is needed at all in modern-day cells. Sometimes it seems that cells could happily get by without it, just exploiting DNA and proteins. It might be that RNA is merely a remnant of the dim and distant past when there were no proteins or DNA in the world.

However, we are not that worried about the particular chemistry of the polymers. What is more important is the very fact of the chain structure. Because all the monomers line up in a chain, we end up with a horrendous number of possibilities, m^N, which cannot all be tried out. This, together with the assumption that a significant proportion of all the primary structures are catalytically active, gives us the following consistent picture.

Once upon a time, some random polymer chains were produced. Their number was negligible compared to the total number of possible combinations. However, in absolute terms it was probably quite a lot. Then there was a selection among the chains. Only the best fitted survived, that is, those that won the battle for the limited stock of the "monomer food." As a result, a very interesting system was formed. Perhaps it cannot quite be regarded as alive yet, but it is certainly worth considering in detail.

11.7 — Spontaneous Symmetry Breaking and the Memorizing of a Random Choice

Let's think again of the picture we have just described. Although the best-fitted "leader" is selected, some spontaneous polymerization can still go on. It may even happen that a chain synthesized spontaneously has a better catalytic ability than the leader. In fact, the probability of such an event is of order 1. Moreover, this may happen at any moment, even right now!

In connection with this idea, a really wonderful letter is often quoted from Darwin to his friend, colleague, and life-time correspondent J. D. Hooker, in which Darwin imagines a warm pond with the necessary combination of salts, sources of light and electricity, and so on, and then raises the question: Why are not new substances capable in further transformations of growing complexity appearing there every day? Why not today? The answer Darwin gave is simple: As there are living creatures everywhere on Earth today, a new chain of a biopolymer, even if synthesized accidentally, is bound to be eaten well before anything interesting can happen to it.

This is exactly what we can say about the system of catalytic chains with a well-defined leader. The leader is represented by a great number of chains. Hence, there is little doubt that the greedy crowd will leave the newly born polymer with no "monomer food," even though it might be a better catalyst.

This situation is a bit similar to the simple mechanical model shown in Figure 11.2. The first diagram, Figure 11.2*a*, depicts a perfectly symmetric yet unstable system. Then the symmetry is broken at random (or spontaneously, as they often say). As a result, the system comes to a stable position (Figure 11.2*b*). The stability means that the random choice (change) that has been made is now "engraved" into the system's memory.

Thus, when the leader shows up in a system of self-catalytic (or mutually catalytic) polymers, we can regard this as the spontaneous breaking of the symmetry. (Indeed, before the leader appeared, all the chains had roughly the same catalytic ability. So we can say that the picture was symmetric.) The random change (choice) is memorized.

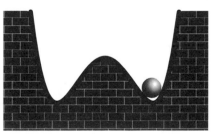

a   *b*

FIGURE 11.2
A mechanical illustration of how a random choice is memorized.

The memorizing of the random choice turns out to be a very interesting thing. Following are a few examples.

11.8 — Right- and Left-Handed Symmetry in Nature

Most people have their hearts not in the middle of their bodies, but on the left-hand side. In contrast, a DNA double helix, a treble helix of collagen, and α-spirals of globular proteins all have a right-handed structure. People drive on the right in most countries, though there are exceptions, such as England, Ireland, and Japan. In engineering, except in some very special cases like the left-hand pedal of a bike, only right-handed screws are used. Why?

Let's start with engineering. A left-handed thread is no worse than a right-handed one. However, imagine a child's do-it-yourself kit where right-handed screws and nuts are mixed up with left-handed ones. The symmetry is, therefore, preserved, but it is quite awkward to play with, to put it mildly! This awkwardness comes from the instability of the symmetrical state. The direction in which the symmetry was broken (right-handed screws) was chosen more or less by accident in the past. However, now it is well established and is retained by standards and tradition. In other words, this choice has been memorized, and the system is now fairly stable. It is quite unthinkable that left-handed screws will suddenly come into fashion!

The situation with molecular "screws" in nature is a little more complex. In atomic nuclei, there are mirror-asymmetric interactions that are called "weak." They are weak indeed. At least, they hardly affect the properties of the atoms. Some time ago, any influence was disputed all together. However, in the 1970s scientists managed to detect it, in very refined optical experiments with bismuth vapor. Weak interactions make right-handed and left-handed spiral molecules differ slightly in energy. This difference is very small; it is estimated to be about 10^{-17} of the characteristic energy. We personally subscribe to the view that such a minute discrepancy could not have played any role, and the modern biosphere is asymmetric in the way it is entirely by chance. The leader-biopolymer, when it first appeared in the primeval soup, happened to have a certain direction of the spiral. This random direction was memorized because the leader was the most successful in reproducing its own copies.

There is a curious fictional story on this subject, about a shipwreck. The victims die of starvation on an unknown island, although beautiful fruits grow there in abundance. The clue is that this island is really a land behind the looking glass. It is a country of left-handed DNA, collagen, and α-spirals. Thus the fruits growing there are no good for eating. The tragic end of this story confirms: don't mess around with a leader! (The leader obviously was the system of biopolymers with spirals in the "wrong" direction.)

(And why is this story about an island? Isn't it because all the countries with left-handed roads are on islands?)

11.9 — The Memorizing of a Random Choice, the Creation of New Information, and Creativity

The appearance and progress of life is not the only case of evolution. There are other examples, such as the emergence and development of languages, literature, art, science, and even the game of chess. All these systems, in a way, do the same things as living beings. You could say that scientific ideas and pieces of literature are also born. Some of them die (but not all!). Many leave offspring. (For example, the first derivation of the heat conduction equation was based on the belief that heat is carried by some sort of medium or ether. Thus, the heat conduction equation is an offspring of a now-dead concept. Similarly, Don Quixote may not have appeared without a whole bunch of mediaeval stories about knights, now happily forgotten.) What is most important for us is that all such systems develop by memorizing random choices. For instance, it is more or less by accident that we use the word "number" for the concept of number. We could have had another word instead (as, say, the Russians do; they call it *chislo*.) However, once made, the random choice is cemented in books and people's memories. The attachment of the words to their meanings is no less stable than the position of the ball in the little hole in Figure 11.2*b*.

As the American biophysicist G. Castler showed in 1962, when a system memorizes a random choice, it creates a novel bit of information. This information is about facts, which were never known or questioned, or even existed before. For example, in which hole is the ball in Figure 11.2*b*? What sort of primary structure does the oxygen-carrying protein have in human red blood cells? Which words are used for one or another concept? How exactly did Tom Sawyer manage to get the fence painted? And so on. As a matter of fact, the memorizing of random choices is also relevant to creativity, both in art and science. If you are interested in further discussion on this, we recommend the enthralling, yet not always uncontroversial, books by M. V. Volkenstein [5].

What about other evolving systems, such as languages? It is interesting that in language there are some general laws that do not depend on random choices, that is, on the meanings that particular words happen to have. Such are the laws of poetry, for example. This becomes clear from the following absurd yet "proper" poem:

Hunkle, chinkle, mrony phar,
Brough I junder whow mee dar?
Up above the fye bo clar,
Hunkle, chinkle, lubby phar.

Is there anything similar in the physics of biopolymers? In the past, certain primary structures were picked at random, out of a huge crowd of "candidates," to be given one or another job to do. Are there any general laws that are not affected by such random choices? There certainly are! They control the formation of knots in DNA (Section 2.6), the hydrophobic-hydrophilic separation of a globular protein (Section 4.7), and many other processes. And many such laws may still be unknown.

11.10 — Conclusion: What Is Still Unclear?

From all we have said we can draw the following conclusion. There are systems where you cannot physically try out all the possibilities because there are just far too many of them. The only way such systems can evolve is by memorizing random choices. This is how new bits of information are created. The first in history as well as the simplest event of this sort was probably the synthesis of polymer chains in the primordial soup.

The very fact that scientists can now talk about such things is exciting. However, it is an even greater achievement that we can realize, discern, and set down, in a precise and clear way, the questions that are still to be answered. Here are just a few:

- Which of the randomly synthesized polymer chains have the capability of self-organization (Section 8.13) and catalytic ability? What are these abilities in particular? How well expressed are they, and with what probability?

- What are the general properties of biopolymers that are independent of their specific functions?

- How were droplets formed in which the concentration of primordial polymers was high enough for self-catalysis to occur? (A question raised by the Russian biochemist A. I. Oparin, who was one of the first to study evolution before life as early as in 1924.)

- At what stage and how did membranes emerge (Section 4.2)?

- In what sort of units could macromolecules have come together, at the next level of structural hierarchy? For example, they could have got into small groups forming spatial aggregates (just like monomers lined up into complex three-dimensional chains at the previous stage). Or, the macromolecules could have organized some sort of "collective farm" engaging in the common cycles of metabolism.

- Is the genetic code just the result of memorizing random choices? Or are those scientists right who think that the genetic code reflects the stereochemistry of the main biopolymers?

As you can see, all the questions are to do with polymer physics. This list could be extended even farther. If you are interested in all this, we strongly recommend the wonderful, thought-provoking book by F. Dyson [11].

There is another fascinating aspect of the origin of life: Is there life elsewhere in the Universe? It might be worth searching for the signs of primordial polymerization even on Mars. Interestingly, this suggestion is now less fantasy and more real science, since it has been proven that some meteorites falling to the Earth are actually coming from Mars (being boosted out of orbit by the powerful impact of a comet or asteriod). Will we soon know more about life in the Universe? Whatever the answer about Mars, likely the most promising places in the solar system are the methane seas on Titan, a satellite of Saturn. Unfortunately, we haven't the foggiest idea how to get there to check! Thus, as it is often said in such cases, we leave this as an exercise for the reader.

To conclude, we would like to warn you that perhaps not all the experts would agree with what we have said in this chapter. Some may argue that whenever physicists try to discuss such matters it is merely naive and useless. One could respond to this with the following words from the famous *Feynman Lectures on Physics*:

> The most important hypothesis in all of biology... is that *everything that animals do, atoms do*. In other words, *there is nothing that living things do that cannot be understood from the point of view that they are made of atoms acting according to the laws of physics*. This was not known from the beginning: it took some experimenting and theorizing to suggest this hypothesis, but now it is accepted, and it is the most useful theory for producing new ideas in the field of biology.... Certainly no subject or field is making more progress on so many fronts at the present moment,[4] than biology, and if we were to

---

[4] It was written in 1963, and it is equally fair now, in 1996.

name the most powerful assumption of all, which leads one on and on in an attempt to understand life, it is that *all things are made of atoms*, and that everything that living cells do can be understood in terms of the jigglings and wigglings of atoms.[5]

Why do some people refuse to discuss the possibility of any physics behind the problem of the origin of life? And believe that life occurred entirely due to a miracle, that is, an event of a ludicrously small probability 10^{-260}, or even smaller? Perhaps, it is merely cowardice, the fear of a very difficult task. No doubt, the task really is extremely complicated. Still, maybe someone will one day dare to try?

---

[5] *Feynman Lectures on Physics*, Vol. 1, © California Institute of Technology. Reprinted by permission of Addison-Wesley Longman, Inc.

Application Polymer Program on the CD ROM[1]

A.1 — Introduction

The *Application Polymer* program is a toy model of a simple world. We suggest that you play with it! To play, you can either watch the movies that we have prepared for you or create new ones of your own. The world that we are suggesting you play with consists of rather simple atoms, much like Democritus of ancient Greece could have imagined. If we define species of those atoms along with a few parameters to describe their properties, the system can mimic a variety of phenomena of the real molecular world.

We suggest that you first watch our movies: This does not take any effort. If you like them, you can try to be creative. This may take some patience and effort,

If, in some cataclysm, all of scientific knowledge were to be destroyed, and only one sentence passed on to the next generations of creatures, what statement would contain the most information in the fewest words? I believe it is the *atomic hypothesis* (or the atomic *fact*, or whatever you wish to call it) that *all things are made of atoms—little particles that move around in perpetual motion, attracting each other when they are a little distance apart, but repelling upon being squeezed into one another.*

R. Feynman, R. Leighton, M. Sands, *The Feynman Lectures on Physics*[2]

---

[1] This appendix was written jointly with Dr. S. V. Buldyrev.
[2] Vol. 1, © California Institute of Technology. Reprinted by permission of Addison-Wesley Longman, Inc.

but it does not require knowledge beyond that of most high school students. (At least, what we are sure they *should* have!)

This appendix is organized as follows. We first describe our simplest movies—those that are even simpler than the ones included with the main text of the book. Then we go a little deeper and explain how our toy world works: what atoms look like, how they combine into molecules, and how they interact and move. After that, we specify in more detail how the properties of a particular system of atoms are encoded for the program to work. Finally, we give some specific suggestions for creating a good movie.[3]

 A.2 — Simplest Movies

 Our first movie is called *"Gas."* It shows an ordinary simple system of identical molecules. What you see here is the chaotic motion of all the molecules. This is called thermal motion or, if you prefer, Brownian motion.

Note that in this first movie all of the molecules are of the same kind; they make up a pure substance, like pure water. Molecules never penetrate each other, meaning that there are forces that repel molecules from each other when they come close (which, of course, happens from time to time). In this sense, the gas is not ideal. One can "measure" its pressure and see that it deviates from Boyle's law. At the same time, it is hard to tell if there are any forces of attraction. But if you look carefully, you can see a slight manifestation of the attractive forces: When two molecules happen to be relatively slow (some of them are always slower than others) and two slow ones meet, they remain coupled for quite some time. This suggests that something interesting can happen at lower temperatures, where all of the molecules move more slowly.

 Our second movie, *"Condensation,"* is just a little bit more complex than the first one. We show the same molecules as in the first movie, except now the temperature is lower. What happens to familiar liquids when you heat them up? What happens to water in your coffeemaker or kettle when you are preparing your breakfast? Yes, of course; it boils, and vapor comes off. Accordingly, if you take a vapor and cool it down, you expect to see condensation. You can see in this movie how this condensation

---

[3] Apart from the movies generated using *Application Polymer*, there is one more movie, #16, called *"Fractal Growth,"* which is generated independently.

looks at the level of molecules. When the temperature goes below the freezing point, the little crystals start to grow out of the liquid. You can see this process in the third movie, called *"Crystallization."*

Thus, forces of attraction are directly responsible for phase transitions, such as condensation and the appearance of liquids. However, this was a pure substance. What can happen if we mix two different substances?

"Phase Segregation" **is our fourth movie, and it is again about a simple nonpolymeric liquid. This time we show what happens if we try to mix two liquids, such as water and oil, that tend not to mix. You can see that the different types of molecule do indeed get segregated. Not perfectly, because there is some diffusion of one component into the other, but most of the molecules segregate very well.**

How did we do that? We took two types of molecules very similar in size that interact in a peculiar way: Similar molecules attract each other, different ones repel each other. Not surprisingly, they don't tend to mix.

We hope that these three examples are sufficient to give you an idea of the richness of our toy world. Much more complex movies, related to the life of this same toy world, are referenced in the main text of the book, and you can certainly watch them again. In the next sections, we shall explain how all these movies are created.

A.3 — Toy Atoms and Toy Molecules

The atoms of the toy world are toy atoms. They are much like billiard balls, but not exactly. The atom shown in Figure A.1 has a hard core (shown dark in the figure), which is indeed exactly like a billiard ball. It is this core that is shown in the movies.

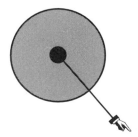

FIGURE A.1
A sketch of an "atom" in the toy world. The dark core represents the region where other atoms can never penetrate, the lighter region represents the region of attraction, and an "arm" can be used to combine with other atoms to form molecules.

FIGURE A.2
The interaction
potential as a function
of distance (*a*) for real
atoms and (*b*) for toy
atoms.

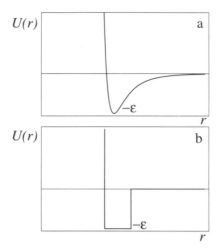

Other atoms can never penetrate there. The lighter area around the core in
Figure A.1 is the region where other atoms feel attracted to this atom, if they
come this close. Remember that normally there are interactions between atoms,
molecules, and other particles in the real world. (A typical potential as a function
of the distance is shown in Figure 7.1 in the main text.) In Figure A.2, we repeat
the real-world potential of Figure 7.1 and show also the toy atoms' interaction
potential. Finally, each toy atom has an "arm," which serves to combine atoms
into molecules. Each arm can hold one (and only one) partner. For example, if
the arm of atom *A* holds *B*, the arm of *B* holds *C*, and the arm of *C* holds *C* itself
we get a three-atom molecule for which the "chemical formula" looks like *ABC*.
This is shown schematically in Figure A.3. Thus, to model molecules we must
specify from the very beginning the entire list of connections. That is, for each
atom, we have to specify which partner will be connected to this atom's arm.

FIGURE A.3
A sketch of a
three-atom "molecule"
in the toy world. Arms
of each atom are not
shown to scale, but are
exaggerated.

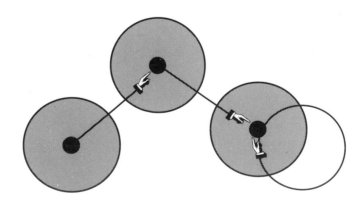

What do the toy atoms and toy molecules represent? What do we try to model and show with them? Certainly, toy atoms do NOT necessarily pretend to model real atoms. They may as well represent much larger "clumps" of material. For instance, when we model polymer chains, we represent each monomer as one toy atom. Clearly, in this case each toy atom plays the role of a pretty big molecule. As a matter of fact, however, they may represent even bigger particles, such as seeds of dust or even viruses: these large particles, when suspended in liquid, can undergo various transformations that can be modeled well using our toy world. Thus, we will keep speaking about atoms and molecules, but the reader must keep in mind that they can mimic many different situations of the real world.

A.4 ─ Physics of the Toy World

A.4.1 Space and Time in the Toy World Because our toy atoms can represent a variety of different real systems, the absolute values of the atoms' sizes lack any meaning. Similarly, absolute values of any time intervals are meaningless. The only relevant figures are those representing *ratios* of different length and time scales.

We do not indicate anywhere any specific characteristic times for the processes shown in the movies. In fact, one can use this toy world system to address interesting questions related to time. For example, we can *compare* times required for some process (for example, collapse) for polymers with, say, 50 and 100 monomers and thus we can obtain a good idea of how this time depends on the chain length.

The situation is a little different for the lengths. A computer cannot simulate an infinite system. Therefore, one length scale is involved in all simulations: it is the size of the box in which all toy atoms and molecules are confined—the toy universe. We shall define all the distances, radii, etc., using an artificial unitary length, which we choose such that the size of the toy universe equals 384 (unless specified otherwise). When you view a movie, you see a box on the screen of your computer, and the size of this box is 384 (for most of the movies).[4]

The toy universe also has very peculiar properties as regards its edges. What happens when, say, a toy atom approaches the lower boundary of the box? The

---

[4] Of course, the number 384 seems like a pretty strange choice; where does it come from? You can easily guess that there is nothing behind this choice but historical coincidence; in this sense, by the way, the toy world is also similar to the real one, where almost all the units are just the fruits of history. This is an excellent example of memorizing a random choice! (See Section 11.7.)

geometry of the toy world is defined in terms of the so-called periodic boundary conditions: this means that an atom approaching a boundary does not "see" anything special, but goes directly to the opposite boundary. In watching the movies, you will see numerous cases where atoms seem to disappear behind the frame. Don't worry, they will not be lost: if an atom goes down to the lower boundary, it appears immediately from the upper one; if it seems to disappear to the right, it appears immediately from the left, etc.

A similar two-dimensional universe can be easily understood as the surface of the torus (like a bagel or a donut): two ways to circle the surface are mapped by going in horizontal and vertical directions on the computer screen.

A.4.2   Dynamics in the Toy World   How do atoms move in the toy world?

They obey standard laws of classical mechanics, Newton's laws. As long as an atom is far enough from all others, no forces act on it, and inertia makes it move in a straight line with constant velocity. Of course, velocities of atoms change upon their collisions: when two atoms come close enough that each enters the region of attraction surrounding the other (see figure A.2), then both of them get faster according to energy and momentum conservation. To be a little more specific, imagine that atoms A and B are colliding. This pair is characterized with the interaction range R_{AB} and interaction strength (or potential energy) ε_{AB}. First of all, the computer has to determine the moment when the two approaching particles become precisely R_{AB} apart; this moment is the beginning of the collision and, given the coordinates and velocities of the atoms at some time, it is predicted as a solution of a simple quadratic equation. Now imagine the line that connects the centers of the two colliding toy atoms at this very moment. For each of the atoms the two components of the velocity vector that are perpendicular to the center-to-center line remain unchanged. The velocity vector components *along* this line change according to the following simple equations:

$$v_A \rightarrow v'_A, v_B \rightarrow v'_B:$$
$$\frac{m_A v_A^2}{2} + \frac{m_B v_B^2}{2} = \frac{m_A {v'_A}^2}{2} + \frac{m_B {v'_B}^2}{2} + \varepsilon_{AB},$$
$$m_A v_A + m_B v_B = m_A v'_A + m_B v'_B, \tag{A.1}$$

where m_A and m_B are the corresponding masses. This yields quadratic equations for both v'_A and v'_B. When ε_{AB} is negative, that is, when atoms A and B attract each other, the toy atoms gain some kinetic energy and become faster (on the average). In some cases ε_{AB} can be positive (when A and B species repel each other), resulting in the inverse effect, namely, that the atoms become slower after entering each other's repulsion area.

Clearly, when two atoms come even closer, and their centers approach the distance r_{AB} and touch each other's hard core, those atoms get reflected perfectly, just like billiard balls. Again, the two components of velocities that are perpendicular to the center-to-center line remain unchanged, while the components along this line change according to the conservation laws (A.1), except this time there is no potential energy ε involved. In this case Eq. (A.1) has one obvious solution: $v_A = v'_A, v_B = v'_B$. This corresponds to the nonphysical situation in which atoms continue to move through each other. However, a quadratic equation always has a second solution, which corresponds to reflection. In the simplest case of equal masses ($m_A = m_B$) this second solution is especially simple—the atoms just exchange their velocities: $v'_A = v_B, v'_B = v_A$.

Similarly, when two atoms are close enough and are within each other's attracting (or repelling) region, but then leave that region, they get slower (or faster). Of course, they still obey the same conservation laws, except this time potential energy ε plays the opposite role (or enters the other side of equation), because particles have potential energy ε_{AB} *before* and don't have it *after*. If the equation Eq. (A.1) does not have real roots—as often happens with quadratic equations—then the atoms cannot and do not exit the potential well, but instead just undergo the "internal reflection."

Finally, similar rules apply for any two atoms that are connected by an arm. An arm acts as an infinitely high potential barrier that does not allow two connected atoms to be farther apart than the arm's length, D_{AB}. More specifically, as long as two connected atoms A and B are closer than D_{AB}, they do not exert any force on each other; as soon as they approach the distance D_{AB}, they undergo "internal reflection," that is, their velocities transform according to the same conservation principles (Eq. A.1), without any potential energy ε involved. There is also a minimal distance d_{AB} for connected atoms, and they undergo elastic reflection if they come closer then d_{AB}. In most cases, $d_{AB} = r_{AB}$, but the program allows these two parameters to be different.

To be honest, we should explain that there are some pathological properties to our toy atoms, which are due to the step-wise character of the potential functions. Indeed, in order to "jump" out of the potential well it is insufficient that the atom have just the appropriate amount of kinetic energy; it must have the velocity with a sufficient component perpendicular to the potential wall. This gives rise to the possibility that even a very fast atom does not jump out of the well, because it has approached the boundary of the well at a small enough angle. In particular, we can artificially prepare two atoms that orbit each other in a quasi-molecule that never breaks apart despite the arbitrarily high speeds of the atoms—because they accurately follow the circumferences of each other's potential wells. In reality, however, it is extremely difficult to realize this truly pathological configuration. In the many-body system it could be created in principle for two atoms upon a very delicate and specific participation of a third (and fourth) partner, but the probability of that occurring is really small, and we have never seen anything like it ourselves.

Thus, to know everything about how all atoms move in the system, the computer only has to solve a couple of simple quadratic equations (Eq. A.1) every time two atoms come close to each other. In fact, it has to solve the equations many times, but still the calculations are fairly simple. This is why even modest computer power is sufficient to follow interesting effects in the toy world.

A.4.3   Past and Future in the Toy World   Pierre Simonde de Laplace (1749–1827) was the first to formulate the views of Newtonian mechanics in the most dramatic way: if we were given coordinates and velocities of *all* particles in the world at some initial moment, we could predict *everything* that happens in the world—in the future as well as in the past. Of course, in the real world there are complications due to quantum mechanics, cosmology, etc. Also, to make such ambitious predictions for the real world, we need a large computer, which itself is part of the world, and thus it must predict its own behavior, and so on. Such speculations lead to deep philosophical discussions, and we shall not consider them here. However, the Laplacian deterministic view seems perfectly workable for our toy world.

Another aspect of the same view is reversibility of the purely mechanical description: if we were able to instantly change the signs of all velocities in the world, then all history should start going backward, repeating precisely step by step in inverse order all the details of the original developments. But this clearly seems to contradict the obvious irreversibility of reality or, as R. Feynman called it, the distinction between past and future.

Thus, our question is about the irreversibility of the toy world. Unlike the real world, in the computer we easily and instantly can inverse all the velocities. Will we be able to "play history" backward? Will counterintuitive events occur, like warming and boiling water in a tea kettle at the expense of heat absorbed spontaneously from the surrounding air? The answer is no: the toy world is irreversible, just like the real one. Past and future in the toy world are distinct in very much the same way as in reality. The source of irreversibility is the rounding of numbers inevitably done by the computer at virtually every step of computation.

A.4.4   Energy and Temperature in the Toy World

In the toy world, just as in the real one, the energy of any closed system is conserved.[5] In particular, the total energy of the entire toy universe, i.e., of the entire box, is conserved—unless we choose to stop running the program and change energy.

On the other hand, we describe various objects of the real world in terms of temperature, and temperature is related to the average kinetic energy of the chaotic thermal motion:[6]

$$\frac{1}{2}T = \left\langle \frac{1}{2}mv_x^2 \right\rangle \tag{A.2}$$

(and similar formulae for v_y in the two-dimensional case, or for v_y and v_z in the three-dimensional case). Here v_x, v_y, and v_z are Cartesian components of the velocity, and $\langle \ldots \rangle$ means average. Usually we deal with bodies that are not isolated but are in contact with other bodies, such as surrounding air molecules, etc. However, in the toy world there is nothing around each body! It turns out that the concept of temperature becomes somewhat tricky for this kind of closed world. The problem is the following: imagine that in the beginning all atoms in the toy world have very slow velocities, such that we would describe it as very cold. Suppose, however, that these small velocities are pointed in such a way that atoms ultimately come closer and enter each other's potential wells. Then, their velocities get considerably faster. Would we then say that it is getting warmer?

This is not at all an artificial example. Rather, it is precisely what is happening in movies such as "*Condensation*" and "*Crystallization*." In the real world condensation and crystallization both are accompanied by the production of a certain amount of latent heat. In precisely the same way, when we take the toy gas shown in the first movie and reduce the speeds of all atoms by some factor we actually make the atoms spend a greater fraction of time close to each other, within the potential wells. On one hand, it *is* condensation, as atoms are closer together on average; on the other hand, energy conservation (Eq. A.1) implies that kinetic energy becomes greater on average at the expense of potential energy, and this latent heat raises temperature. We shall come back to this later, while commenting on specific movies. Here we mention only that in order to really cool down the

---

[5] Potentially, energy conservation could be violated due to the rounding of numbers in the computer; we have checked that this does not happen to a noticeable extent.

[6] Usually, the Boltzman constant, k_B, also appears in Eq. (A.2), because of the traditional units in which temperature is measured. For the toy world, we define temperature by Eq. (A.2) such that it is measured in the same units as energy, so there is no need for a constant.

toy world we must perform several cycles of the reduction of the kinetic energies of the toy atoms.

It must be noticed also that the only meaningful units for energy and temperature in the toy world are related to the interaction potential energies ε. Just as in the cases of length and time, we do not specify any relation of those energies to any specific number of *joules* or *degrees* of the real world, because our toy atoms may mimic very different physical realities.

A.4.5 Which *Physical* Parameters Are to Be Encoded in the Program?
We are ready now to specify all the necessary parameters for the system of toy atoms. We must make a clear distinction between "physical" parameters that control the motion of toy atoms and other parameters that define the way the system appears on the screen. Let us begin with physics. To specify the system, we need to know:

- the particle types, and the mass m for each type; and

- the lengths of the arm connections, ranges, and energies of interactions for all possible pairs of atom species. Specifically, for each pair of atom species A and B we need to specify both maximal D_{AB} and minimal d_{AB} lengths of the arm that can connect them, interaction energy ε_{AB}, interaction range R_{AB}, and hard core repulsion distance r_{AB}.

To specify the screen appearance of the system, we need to know the radius r_A and the color for each atom type.

NOTE: r_{AB} here is the minimal distance between the centers of the particles of types A and B. If the distance between the centers of particles is less than R_{AB}, the particles interact with energy ε_{AB}; otherwise the particles do not interact. Obviously R_{AB} always should be larger than r_{AB}. In almost all our simulations $r_{AB} = d_{AB} = r_A + r_B$, where r_A and r_B are the radii of the corresponding images that appear on the screen. This convention ensures that in the two-dimensional case the atoms move on the screen like billiard balls: at the moment of their collision their visible hard cores touch each other but never intersect. However, it is possible to define r_{AB} and d_{AB} independently of r_A and r_B. In our movie descriptions we always specify when this is the case, assuming by default the billiard ball convention just described.

A.4.6 Why Do We Think Our Toy World is Really Amazing?
When you watch the movies featuring the "events" in our toy world, you may ask yourself: what is so special about these movies? Are they any better than many other

computer-generated movies with a variety of characters and exciting stories? Unlike many other movies, we do not control our characters; they do whatever they do according to their own, actually very simple, rules: all they know is conservation principles (Eq. A.1), and *nothing* else. And yet, despite the primitive and unanimated character of the laws governing the toy world, the world itself appears to be almost unbelievably rich. It is almost like the difference between a child who speaks, albeit with a primitive vocabulary, and a puppet that speaks perfectly, but with an actor behind the scene. Our toy world is truly amazing precisely because there is nobody "behind" it!

The Individual Movies

A.5.1   MOVIE 1: *GAS*   This movie is mentioned in Section A.2.

Description of the System

Two-dimensional system (this is why you will not see anything if you choose the xz or yz projections out of the "View" menu). Number of atoms: 1024. All atoms are of identical type $A = 1$. Hard core radius is $r_A = 3.3$. Distance of attraction is $R_{AA} = 12$; attraction energy is $\varepsilon_{AA} = -1$. Total energy of the system is $E = 1336.9$.

What is Not to Be Missed while Watching this Movie

While thermal motion in the gas is really chaotic, several examples may be noticed here and there when atoms come close to each other and remain together for some period of time. This indicates that attractive forces are indeed present and suggests that condensation is possible upon lowering overall energy.

Interesting Alterations that Can Be Done to this Movie

Stop in the course of the simulation and lower temperature somewhat. Repeat several times.

A.5.2   MOVIE 2: *CONDENSATION*   This movie is mentioned in Section A.2.

Description of the System

The same as in the previous movie, "*Gas,*" except overall energy is lower, $E = -2374.8$.

Not to Be Missed

As explained earlier, temperature increases upon condensation in our system, because there is no heat drainage. This is why the surface of the liquid phase remains rather rugged, and surface tension remains relatively small.

When you look at the gas phase, pay attention to the fact that atoms there often form clusters, or clumps of two or three; usually, they escape from liquid phase this way and then remain together throughout their journey in an almost empty gaseous region. This tendency goes along with the manifestations of local short-range ordering in the liquid phase. Notice small regions in the liquid where as many as a dozen atoms form a relatively regular pattern and maintain it for a considerable time. Clearly, this indicates the possibility of crystallization, but as the energy is still too high, all these ordered clusters are eventually destroyed and short-range order does not proliferate to longer distances.

Interesting Alterations

Stop in the course of simulation and lower energy somewhat; it is possible then to obtain the liquid with a much smoother surface.

A.5.3   MOVIE 3: *CRYSTALLIZATION*   This movie is mentioned in Section A.2.

Description of the System

The same as in the two previous movies, "*Gas*" and "*Condensation*," except energy is even lower, $E = -3465.8$.

Not to Be Missed

To be frank, we did not expect ourselves that the liquid in this simulation would freeze. We kept experimenting with condensation, looking at the liquid at successively lower temperatures. The liquid existed for quite a while on the screen of our computer, and we did not expect anything more interesting to happen. Luckily, we forgot to switch off our computer, and the program kept running overnight. When we looked at the screen in the morning, we had a small shock: the liquid had frozen! It was almost like waking up after a warm rainy night and finding everything frozen and covered with a blanket of snow outside.

There are several interesting features to this movie. First, note that the size of the crystals is growing in a peculiar way: small crystals disappear, while big ones grow at their expense. Further, you'll see that the liquid phase coexists with the crystal solid—in exactly the same manner as ice floating on the water in a lake. Moreover, the gas phase also exists: watch individual atoms or small clumps escaping from the condensed surface from time to time. This means that our system is close to the triple point. (Remember that triple point is the set of conditions under which three phases, liquid, solid, and gaseous, can coexist.)

Interesting Alterations

Upon changing the attraction distance from $R_{AA} = 12$ to $R_{AA} = 11$, this system can be made to freeze into another crystal form, with atoms forming a square, rather than a triangular, lattice.

A.5.4 MOVIE 4: *PHASE SEGREGATION* This movie is mentioned in Section A.2.

Description of the System

There are two versions of this movie, two-dimensional ($2d$) and three-dimensional ($3d$). Two types of atoms, A and B, shown in blue and red, respectively. Number of atoms in the two-dimensional case: 512 A atoms and 512 B atoms; overall energy of the system is $E = 1764.2$. In three dimensions there are 500 atoms of each type; total energy of the system is $E = 80$; size of the universe is 225. Other parameters are shown in Table A.1 for both two-dimensional (labeled $2d$) and three-dimensional ($3d$) models.

		A $m = 1$	B $m = 1$
A	2d:	$r = 5, R = 20, \varepsilon = 0$	$R = 20, \varepsilon = 1$
	3d:	$r = 10, R = 22, \varepsilon = 1$	$R = 24, \varepsilon = -1$
B	2d:	$R = 20, \varepsilon = 1$	$r = 5, R = 20, \varepsilon = 1$
	3d:	$R = 24, \varepsilon = -1$	$r = 10, R = 22, \varepsilon = 1$

TABLE A.1
Parameters for Movie 4,
"Phase Segregation."

Not to Be Missed

As the system energy is low enough overall, its state with mixed A and B components is absolutely unstable. What is happening in such a system is called spinodal decomposition. Watch coarsening of the red and blue islands and gradual decreasing of the interfacial surfaces. In the two-dimensional version of the movie, you will see *two* regions of red and *two* regions of blue formed as the strips on the screen. It will look like nothing else is going to happen, but don't give up, keep watching! Energetically, it is more favorable to decrease the interfacial surface even further, that is, to gather all molecules of each type in just a single phase. It is, however, difficult for the system to get to this energetically favorable configuration, and thus the state with two stripes appears to be a rather long-living, metastable one. Eventually, however, fluctuations of two opposite surfaces touch each other, and the last step of coarsening begins. In the end, you will see indeed one red phase and one blue phase. (To fully appreciate that you have to remember about periodic boundary conditions: blue regions on the left and on the right are *the same*!) Notice also that even in the system with very good phase segregation there are still a certain number of blue molecules that keep diffusing through the red phase, and vice versa.

Interesting Alterations

As you can see, we choose quite different and arbitrary parameters in the two-dimensional and three-dimensional cases. Try your own values and develop an intuition of how these parameters, including temperature, affect the behavior of the system and the speed of the process.

A.5.5   MOVIE 5: *FLEXIBILITY*   This movie is mentioned in Section 2.2.

Description of the System

Three-dimensional system consisting of 99 identical atoms connected into one chain. Hard core radius of each atom is $r_A = 1.9$. Arm maximal length for every connection between atoms is $D_{AA} = 5.36$. No attraction forces ($\varepsilon = 0$).

In the beginning, the entire chain is stretched along the x axis. Then, one central atom, #50, is assigned velocity 0.98 in the y direction while all others are assigned 98 times smaller velocities in the opposite direction (to keep overall momentum zero). Similarly, atom #50 is assigned velocity -1 in the z direction and atoms #49 and #51 are assigned velocities 0.5 in the z direction.

Not to Be Missed

As this system is three-dimensional, it is instructive to view it in different projections. As the initial push in the z direction of atom #50 is compensated by the opposite pushes of the neighboring atoms, the development is much faster and stronger initially in the xy plane than in the xz plane. Watch how tails gradually are getting involved in the motion. Eventually, you will see that the whole polymer is coiled up and forms a quite isotropic shape.

Interesting Alterations

Try to change initial conditions, such that the push is given in one y direction only. Then, as there are no solvent molecules outside, there will be no way for the system to leave the xy plane and it will remain two-dimensional forever.

A.5.6   MOVIE 6: *SOAP*   This movie is mentioned in Section 4.2.

Description of the System

This system is among the most complex in our movies, although it is two-dimensional. There are three components to the system: dirt (oil), water, and a soap (or surfactant). Water molecules are shown as blue (or light blue on some screens), structureless "atoms." Oil "atoms" are shown red; each oil molecule consists of four atoms connected linearly (like *OOOO*). Each soap molecule include one hydrophobic atom, shown dark blue, and one hydrophilic atom, shown purple. Parameters of the system are listed in Table A.2. There are 48 oil molecules, 192 soap molecules, and 448 water molecules in the system. Overall energy of the system is $E = 347.2$.

TABLE A.2
Parameters for Movie 6, "Soap."

	water lt. blue $r = 5$ $m = 1$	oil red $r = 5$ $m = 1$	soap: phob. dk. blue $r = 5$ $m = 1$	soap: phyl. purple $r = 5$ $m = 1$
water	$\varepsilon = 0$	$\varepsilon = +1$ $R = 14$	$\varepsilon = +2$ $R = 20$	$\varepsilon = -2$ $R = 20$
oil	$\varepsilon = +1$ $R = 14$	$\varepsilon = -1$ $R = 14$ $D = 14$	$\varepsilon = -2$ $R = 20$	$\varepsilon = +2$ $R = 20$
soap: phob.	$\varepsilon = +2$ $R = 20$	$\varepsilon = -2$ $R = 20$	$\varepsilon = 0$	$\varepsilon = 0$ $D = 14$
soap: phyl.	$\varepsilon = -2$ $R = 20$	$\varepsilon = +2$ $R = 20$	$\varepsilon = 0$ $D = 14$	$\varepsilon = 0$

Not to Be Missed

Oil originally forms a big, dense drop in water. Without soap, water could never penetrate this drop, because the oil molecules strongly attract each other, with

$\varepsilon = -1$. Now soap comes into play. The dark blue parts of the soap molecules attract very strongly to the oil, while the purple parts of the soap attract very strongly to water. As a result, soap molecules pull the oil drop apart, covering each small droplet of oil and shielding it from water. Notice also the small vesicles that the soap molecules sometimes form in the water without any oil. These are soap bubbles!

Interesting Alterations

What would happen if soap lost its strength? Save the last configuration where the oil clamp is completely broken apart and put zero interaction energies for the soap atoms. What will happen if the initial temperature of the system is reduced? Is it better or worse for cleaning? Is it possible to make soap stronger so it will work even in cold water? Try changing the interaction energies: make the interaction energy between the hydrophobic atom of soap and water equal to 0.

A.5.7   MOVIE 7: *MICELLES*   This movie is mentioned in Section 4.2.

Description of the System

This two-dimensional system consists of water, represented by purple atoms, and surfactant molecules, which consist of two atoms each, one red hydrophobic atom and one blue hydrophilic atom. There are 200 water atoms in the system, and 50 surfactant molecules. The energy of the system is -6. Other parameters are presented in Table A.3.

TABLE A.3
Parameters for Movie 7, "Micelles."

		water $r = 9$ $m = 1$	hydrophobic $r = 3$ $m = 1$	hydrophilic $r = 9$ $m = 1$
water		$\varepsilon = 0$	$\varepsilon = +1$ $R = 32$	$\varepsilon = -1$ $R = 32$
hydrophobic		$\varepsilon = +1$ $R = 32$	$\varepsilon = -1$ $R = 32$	$\varepsilon = +1$ $R = 32$ $D = 32$
hydrophilic		$\varepsilon = -1$ $R = 32$	$\varepsilon = +1$ $R = 32$ $D = 32$	$\varepsilon = 0$

Not to Be Missed

Formation of micelles is seen very clearly. Note that the micelles are not frozen but keep exchanging particles. Large micelles are more stable than small ones, but even large micelles from time to time lose some molecules; such molecules then start to wander around in the solution, and the solution therefore always contains some amount of diffusing single molecules. Note that the state of micelles can be reasonably described in terms of temperature, because the surrounding water plays the role of a thermal bath, such that the temperature of the solution does not change much in the course of the simulation.

Interesting Alterations

Change the interactions, increase temperature, and increase surfactant concentration.

A.5.8 MOVIE 8: *BROWNIAN MOTION*   This movie is mentioned in Section 5.2.

Description of the System

Almost the same system as in movies 1–3, with 1024 atoms of mass 1 each. White spheres represent atoms that have hard core distance $r_{AA} = 6.6$ and attractive distance $R_{AA} = 12$ with energy $\varepsilon = -1$. One additional particle added in the system, shown blue, has mass 100 and visible hard core radius $r_B = 12$. It repulses the atoms from the hard core distance $r_{AB} = 14.29$. The energy of their attraction ε_{AB} is 0. For simplicity, regular atoms are shown as points with visible radii $r_A = 0$. Total energy of the system is $E = 1191.4$.

Not to Be Missed

It is interesting that sometimes the big blue walker moves rather regularly in one direction, but eventually always changes course and keeps wandering without any regular bias.

Interesting Alterations

Measure mean square displacement of the walker and watch the square root law (see Section 5.4) at work.

A.5.9   MOVIE 9: *POLYMER SOLUTION*   This movie is mentioned in Section 5.6.

Description of the System

Three-dimensional system, all atoms are identical, with mass 1, hard core radius $r = 5.9$, and without any attraction potential, $\varepsilon = 0$. Atoms are connected into chains: each chain has 32 atoms, with length of connecting arms $D = 16.6$. In the first part of the movie, representing dilute solution, there are 20 chains in the solution. In the second part of the movie, showing semi-dilute solution, there are 64 chains. Chains are shown in different colors only to help the viewer differentiate them.

Not to Be Missed

First, look at the dilute solution. It may seem that the chains tangle up in each other, but this is only an illusion, due to the fact that the chains project on top of one another. Look carefully at the different projections to check that the chains are indeed almost independent.

Second, view the semi-dilute regime. There is still a lot of empty space in the system, as you can see by the presence of black spots on each projection; clearly, black spots mean long holes throughout the system in the corresponding direction. Thus, this solution is far from being concentrated. Nevertheless, looking at different projections, you can see that chains do tangle with each other and behave in this sense interdependently.

Interesting Alterations

Go to even higher concentrations to see stronger entanglements. There is actually still considerable room to increase concentration in the semi-dilute regime, because even with 64 chains the system is fairly close to the crossover threshold concentration c^*. The problem is the computational power required for larger numbers of chains, so try this only if your computer is fast enough.

A.5.10   MOVIE 10: *SWELLING*   This movie is mentioned in Section 7.2.

Description of the System

Chain, shown in three dimensions, consists of 1000 identical atoms, each of mass $m = 1$ and hard core radius $r = 1$. Connection arm length is $D = 2.82$. Attractive potential $\varepsilon = -1$ acts within radius $R = 2.82$.

Initial energy of the system corresponds to temperature $T = 1.8$. Several independent long runs allowed us to find that this corresponds to θ temperature; that is, at this temperature attraction compensates for the hard core repulsion. During this movie, energy of the system is gradually increased, finally reaching $T = 2.2$.

Not to Be Missed

Notice the appearance of long, relatively stretched, loops and passages of the chain while it is swelling in the "good" solvent. As to the order of magnitude of the effect, the theory predicts the swelling ratio to be about $N^{1/10} = 1000^{1/10} \approx 2$, and this is indeed roughly what we see: the linear dimension of the coil grows by about a factor of 2.

Interesting Alterations

Consider swelling in two dimensions, where the swelling factor is expected to be considerably larger, about $N^{1/4}$ (between 5 and 6 for a 1000-link chain).

A.5.11 MOVIE 11: *COLLAPSE*   This movie is mentioned in Section 8.8.

Description of the System

Same system as in the previous movie, "*Swelling.*" Initial conformation of the chain is swollen, it is prepared at the energy corresponding to the "good" solvent temperature $T = 2.2$ above the θ point (which is about 1.8). During the movie, energy is gradually decreased, such that the corresponding temperature drops to about 1.6, which is below the θ point.

Not to Be Missed

Besides the very fact of collapse, with impressive shrinking of the polymer size, notice also how small beads, or raisins, of a condensed polymer gradually appear all along the polymeric chain.

Interesting Alterations

Study what happens to longer chains, what happens in the case of an abrupt (instead of gradual) temperature drop, and the role of friction (which will appear, of course, if we add a few thousand solvent atoms). All these projects require considerable computational power, so try them if your computer is fast enough.

A.5.12   MOVIE 12: *KNOT*   This movie is mentioned in Section 8.12.

Description of the System

System is three-dimensional.[7] Chain consists of 109 atoms, each of mass $m = 1$ and hard core radius $r = 7.08$. No attractive potentials ($\varepsilon = 0$). Atoms are shown in different colors to help guide the eye. Connecting arm length is $D = 20$. The relation between r and D ensures that parts of the chain can never cross each other (true as long as $D < 2\sqrt{2}r$). Initial configuration of the chain is prepared in the shape of the simplest nontrivial knot, called trefoil.

Part I of the movie features a ring polymer (arm of atom i holds atom $i + 1$ for all i from 1 to $N - 1 = 108$ and arm of the atom $N = 109$ holds atom 1). Part II shows a linear (open) polymer, where arm of the terminal atom 109 holds the atom 109 itself.

Not to Be Missed

The knot in Part I can never be unknotted (because $D < 2\sqrt{2}r$) and the chain remains with unchanged topology no matter long we watch it. Notice that the typical configurations of the chain are relatively compact.

The open polymer in Part II has an open end and thus the knot resembles more what one would call a knot in everyday life. This knot can be, and eventually is, untangled. Notice, however, how long the process takes. As soon as the knot untangles itself, or "jumps away from the chain end," the chain becomes much less compact and its size grows considerably. To understand this movie, it is important to remember about periodic boundary conditions: when the chain goes to the screen edge, its continuation is always seen at the opposite edge.

Interesting Alterations

There are many interesting games to play with knots. You can try longer chains and study the dependence of the untangling time on the chain length. Also, you can make more complex knots and study the dependence of the chain size on knot complexity. If you incorporate attractive potential, knots can be examined in the shrunken polymer chain. Again, all these projects require a pretty powerful computer.

---

[7] Lines can tangle in three dimensions only: in two dimensions they cannot intersect; in four dimensions every entanglement and every knot can be untangled.

A.5.13   MOVIE 13: *LINK*   This movie is mentioned in Section 8.12.

Description of the System

System is in many ways similar to the previous one, in the movie "*Knot*." However, there are now two ring chains, of 40 atoms each, and the initial configuration is prepared such that these two rings are entangled and can never break apart. Here $r = 9$ and $D = 25.2$, but still $D < 2\sqrt{2}r$.

Not to Be Missed

As one would expect, the two rings always remain together. Notice how long the symmetry of the initial configuration is preserved. Note also that the conformations are relatively compact. As in the case of knot, topological constraints and restrictions lead to an increase of polymeric entropic elasticity, which provides greater resistance to the excluded volume effect, thus not allowing the entangled system to swell to the extent it would if there were no constraints.

Interesting Alterations

Modify the system slightly by opening at least one of the rings (similar to Part II of the previous movie, "*Knot*"). We expect (though we did not try it), that after some waiting time the two molecules will find a way to break apart and diffuse away from each other. Another suggestion is to reduce the hard core radii of the atoms just a bit, such that the condition $D < 2\sqrt{2}r$ is violated.

A.5.14   MOVIE 14: *FOLDING*   This movie is mentioned in Section 8.14.

Description of the System

One three-dimensional polymer chain of 50 atoms immersed in a solvent of 943 atoms. Solvent atoms are shown as points (otherwise nothing would be seen on the screen but solvent). All atoms (both solvent and monomers) have mass $m = 1$ and hard core distance $r_{AA} = 18$. Connecting polymeric arms are of the length $D = 28$. There are no attractive potentials between solvent atoms or solvent atoms and monomers. For monomer-monomer interactions, the range is $R = 28$, and interaction energies range from $\varepsilon = -0.9$ to $\varepsilon = +0.9$. This means that there are 10 different kinds of monomers, with different interaction energies. Some attract each other, some repel. Corresponding atoms are colored according to the rainbow, with red end colors marking strong attractors and blue-violet end colors marking repulsors.

The polymer behavior in this movie can be well described in terms of temperature, as a considerable amount of solvent, with large enough heat capacity, plays the role of a thermal bath. In the beginning, temperature is $T_0 = 1$. At some moment in the middle of the movie the temperature is instantly reduced[8] to $T_1 = 0.12$. The moment of temperature quench is easy to notice by the sudden drop of intensity of the solvent atoms' thermal motion.

Not to Be Missed

Nothing very remarkable seems to be happening in the system as long as it remains at high temperature in the swollen coil-type state. Upon temperature quench, the polymer collapses pretty fast and acquires the conformation in which red and yellow monomers are preferably located inside the globule, forming a "hydrophobic core," while blue and violet monomers mostly are closer to the surface of the globule, forming a "hydrophilic fringe."

Interesting Alterations

This is one of the richest systems, with almost an inexhaustible set of possibilities: change the scheme of interactions, the number of monomeric species, their sequence along the polymer, etc.

A.5.15 MOVIE 15: *REPTATION* This movie is mentioned in Section 9.3.

Description of the System

Three-dimensional system, consisting of 121 "heavy" chains, with 32 monomers in each, and one "light" chain of the same length, shown blue. All atoms have hard core distance $r_{AB} = d_{AB} = 11.8$; there are no attractive potentials. Connecting arms in the polymers are all of the same length, $D = 16.6$. Heavy atoms have masses $m = 10000$ each, each light atom has mass $m = 1$. The visible radii of the heavy atoms are shown very small in order not to obscure the view of the light chain motion.

Not to Be Missed

With very large masses, heavy chains almost do not move, and thus the light blue chain moves in a medium of essentially immobile obstacles. By choosing

---

[8] Computationally, reduction of temperatures can be achieved by instant redefinition of all velocities by a certain factor, as this rescaling does not affect the shape of Maxwellian distribution.

the appropriate projection, it is possible to see that the blue chain moves in a snake-like fashion, which is called reptation.

Interesting Alterations

It is interesting, though challenging, to prepare an even denser matrix of heavy chains, where the reptation character of the motion should be more pronounced. You can also make a longer light chain, and generally examine the chain length dependence of the reptation time.

[*There is also Movie 16, "Fractal Growth," mentioned in Section 10.8, which is generated completely independently of the Application Polymer.*]

How to Play with the *Application Polymer*

The purpose of the *Application Polymer* is twofold. First, we hope that watching the movies we have prepared will be enjoyable and will promote intuition. Second, the application potentially can be employed as a scientific instrument to examine and answer various questions. Accordingly, one can use *Application Polymer* in two different regimes:

1. Playing movie files (files with names *.mov).
2. Doing actual simulation for a system of toy atoms according to the rules of the toy world discussed above.

In the first regime, your computer does not do any calculations except reading the coordinates of atoms from a movie file. The computer displays frame after frame, and shows atoms and connecting arms (bonds). Atoms are displayed as circles of various sizes and colors. The information on how to display atoms is in the movie files, just before the first movie frame. This information is not enough to perform calculations, and you cannot change anything in the movie except to view the system from various directions, change the movie speed (which has nothing to do with the temperature of the system, of course), and rewind and fast-forward the movie.

In the second regime, your computer has to calculate atom coordinates and velocities according to Eq. (A.1). It must solve an enormous number of quadratic equations and sort the collision times in the order in which they happen.

As in a real experiment, the result of a computer experiment depends on the "experimental" setup, that is, the properties of the toy atoms and molecules, the initial conditions of the system (the coordinates and velocities of all atoms), and the actions of the user during the experiment (for example, reducing or increasing the kinetic energy of the system). All the necessary information to start the computer experiment is stored in the text files *.mtx, which can be written or modified by the user using a regular word processor. The exact rules for how to modify the text files are presented in the tutorial text file manual.mxt and many sample text files on the CD-ROM.

To continue the analogy with a real lab experiment, the computer experiment can also take many hours or even days. It can be successful by producing an interesting phenomenon or result that may or may not have been predicted or expected by the experimenter. It also can be unsuccessful for many various reasons. For example, if the initial text file is designed incorrectly, the application can produce a diagnostic message and quit. Or the number of atoms may be too large to perform calculations on your computer, or it may take your system too long to equilibrate, or the computer can crash in the middle of the calculation due to a power cutoff—and so on. In other words, the whole new art of doing computer experiments is fully analogous to the art of real laboratory experiments, except that you do not need any expensive lab equipment but rather a computer. Here we can make only general recommendations about how to perform computer experiments.

We suggest that you begin with computer experiments employing initial configurations from the text files we have provided. These configurations actually served as initial conditions for the movies that we have recorded on the CD-ROM. We have also included similar text files with a reduced number of atoms for those with less powerful computers. You have to open a file with an extension *.mtx using the *Application Polymer*. It may take a while before the system will appear on the screen. You will see that in most cases (if the number of atoms is larger than one hundred) there are practically no changes on the screen for a long time: Atoms perform only very slow chaotic movements. Don't be disappointed—it sometimes took us several days to record the movies using much more powerful computers than most people can afford at home. If you want to record your own movie, we recommend that you do the following:

1. Reduce significantly the screen update rate, because it takes a large amount of computer power to draw atoms on the screen.

2. Start writing the movie file on the hard disk of your computer. The application will start to record the coordinates of atoms after every specified amount of the toy simulation time. It is not very easy to understand in advance what time interval to use. A general clue is that according to Eq. (A.2), an atom

of unitary mass $m = 1$ at $T = 1$ travels about 1 pixel on the screen in a given direction in a unitary time of our toy universe. Since the atom size is about ten pixels, it is wise to save movie frames every 10 units of time, i.e., when the atom displacement is comparable with its diameter. In general, the frame-saving interval should be inversely proportional to the atom mass and square root of temperature. However, you should always consider the amount of space available on your hard disk.

3. Start writing the parameters of the system, such as temperature, potential energy, radius of gyration of the molecule with the largest mass, and pressure. This information will be recorded as a five-column text file *.par with the toy time of the measurements as the first column. You can use this information for subsequent analysis by standard graphing or spreadsheet software. The recommendations on the time interval between measurements are the same as for the saving of the movie frames.

4. Leave the computer on and do whatever you want (go to bed, have dinner, take a walk, read a book, or watch TV. . .)—Be patient!

5. Check that the application is running OK every several hours[9] and save the current configurations as text files *.mtx. Using these files, you can restart the computer experiment if you have to interrupt it. Moreover, you can modify these *.mtx files the same way you can modify the text files provided by us.

6. Quit the application and restart it opening the movie file you just saved. You will see the development of the system that you have been running for several hours or even days in a matter of a few minutes. For example, you really can see the effect of temperature change during a computer experiment: the atoms will get faster or slower!

The most complicated task of course is to learn how to write the input text files with an initial system configurations. The problem is how to arrange many hard spheres of various sizes and link them together with bonds in the virtual space of our toy universe in such a way that they do not self-intersect. It is tricky! To this end, we developed two kinds of *.mtx files. The first one is the complete description of the system with all atom coordinates and velocities listed. These files are almost impossible for a human being to write from scratch. But it is

---

[9] Of course, we are not urging you get up in the middle of the night and rush to your computer to check what is growing on the screen, but we know that some individuals will do that without our recommendation!

easy to modify these files with a word processor once they have been written by a computer. A second kind of *.mtx files can be used to built the initial but unrealistic configuration in which all atoms are present, do not self intersect, and are properly linked to each other. In these configurations, the system consists of $i \times j \times k$ identical cells organized in perfect rows, columns, and layers, each cell containing equal numbers of different atoms and molecules. Thus, instead of locating the hundreds of atoms of the entire system, one can define only several atoms in one cell. And the other cells will be the perfect copies of this particular cell. But the entire configuration would look like an artificial cubic crystal. What is needed to get random realistic configuration is to raise temperature to a very large value so that all interaction energies are much smaller than T and "boil" the system until it is completely random. Then you can save the system configuration as an *.mtx file of the first kind and make necessary changes to it. Even more complicated recipes can be used to build up a good system. Those who know programming can write a simple program to prepare, say, a linear polymer chain. Others can cut such configurations from one *.mtx file and paste them in another.

It seems to us, based on our own experience, that playing with *Application Polymer* is both entertaining and rewarding. We hope that you will share this view. In the meantime, it does take considerable time, effort, and patience. We wish you success, and hope you will like it!

Suggested Readings

This list is by no means a systematic bibliography on the subject. It includes, in almost random order, both serious monographs and textbooks and simple popular books for a general audience.

[1] de Gennes, P. G., *Scaling Concepts in Polymer Physics*, Ithaca, NY: Cornell Univ. Press, 1979.

[2] Cantor, C. and P. Schimmel, *Biophysical Chemistry*, New York: Freeman, 1980.

[3] Doi, M. and S. F. Edwards, *The Theory of Polymer Dynamics*, Oxford: Oxford University Press, 1986.

[4] Grosberg, A. Yu. and A. R. Khokhlov, *Statistical Physics of Macromolecules*, New York: AIP Press, 1994.

[5] Volkenstein, M. V., *Molecular Biophysics*, New York: Academic Press, 1977; *Physics and Biology*, New York: Academic Press, 1982; *General Biophysics*, New York: Academic Press, 1983; *Physical Approaches to Biological Evolution*, Berlin and New York: Springer-Verlag, 1994.

[6] Eigen, M., and P. Schuster, *The Hypercycle, a Principle of Natural Self-Organization*, Berlin and New York: Springer-Verlag, 1979; M. Eigen and R. Winkler, *Laws of the Game: How the Principles of Nature Govern Chance*, New York: Harper & Row, 1983; M. Eigen and R. Winkler-Oswatitsch, *Steps Towards Life: A Perspective on Evolution*, Oxford–New York: Oxford University Press, 1992.

[7] Trigg, G. L., *Crucial Experiments in Modern Physics*, New York: Van Nostrand Reinhold, 1971; *Landmark Experiments in the Twentieth Century*, New York: Crane Russak, 1975.

[8] Schrödinger, E., *What Is Life? The Physical Aspect of the Living Cell*, Cambridge–New York: University Press, Macmillan, 1945; Enlarged edition, Cambridge-New York: Cambridge University Press, 1992.

[9] Frank-Kamenetskii, M. D., *Unraveling DNA*, New York: VCH, 1993.

[10] Monod, J., *Chance and Necessity: An Essay on the Natural Philosophy of Modern Biology*, New York: Knopf, 1971.

[11] Dyson, F., *Origins of Life*, Cambridge-New York: Cambridge University Press, 1985.

[12] Morawetz, H., *Polymers. The Origins and Growth of a Science*, New York: Wiley, 1985.

[13] Mandelbrot, B., *Fractals: Form, Chance and Dimension*, New York: Freeman, 1977; *The Fractal Geometry of Nature*, New York: Freeman, 1982.

[14] Feynman, R. P., *Character of Physical Law*, Cambridge: MIT Press, 1967.

[15] Weinberg, S., *The First Three Minutes: A Modern View of the Origin of the Universe*, New York: Basic Books, 1988.

90000

9 780123 041302

6 08628 41309 9

ISBN 0-12-304130-9